The SecRet Life of the PERIODIC TABLE

Unlocking the mysteries of all 118 elements

DR BEN STILL

The Secret Life of the PERIODIC TABLE

An Hachette UK Company
www.hachette.co.uk

First published in Great Britain in 2016 by
Cassell, a division of Octopus Publishing Group Ltd
Carmelite House
50 Victoria Embankment
London EC4Y 0DZ
www.octopusbooks.co.uk

Edited and designed by Susanna Geoghegan Gift Publishing

ISBN 978 1 84403 885 5

A CIP catalogue record for this book is available from the British Library.

Printed and bound in China

10 9 8 7 6 5 4 3 2 1

Contents

Introduction

'Humans are pattern-seeking story-telling animals, and we are quite adept at telling stories about patterns, whether they exist or not.'
Michael Shermer

'You're a bright kid but you've got no common sense!'; my mum's catchphrase when I was growing up. I now understand that this is no bad thing for a scientist. Common sense, as my mum puts it, is 'the most likely reason for what we experience'. It is our evolved way of responding to situations automatically, allowing us to gauge the world we live in.

Common sense has evolved with humans, determined by natural selection. Responses which increased the chance of living a long and fruitful life allowed humans time and resources to reproduce. Such responses and ways of thinking

then became ingrained in the next generation. Those who made poor decisions about their environment lived shorter lives and were less likely to contribute to the next generation.

Man versus bear

As an ancient ancestor sleeps they are awoken by a noise. The leaves of a nearby bush are rustling. There is a lot more wind about than bears, so the most likely cause of the disturbance is the wind. Thinking rationally by weighing up the probabilities, they go back to sleep; but what if they are wrong? What if the rustle was actually

Logical ancestors became lunch for bears.

The secret life of the periodic table

caused by a bear? They would likely be eaten by the bear and thus no longer be able to reproduce. On the other hand, if they had assumed the unlikely possibility of the rustling bush being caused by a bear and left their bed to check, then they would have had a greater chance of surviving. The longer an individual survives the more likely they are to reproduce and pass on this way of thinking to the next generation.

The bias towards a pattern that is unlikely, and in the majority of cases turns out to be false, ensures survival. Natural selection therefore favours survival of animals which consistently give weight to illogical patterns in their experience of nature, on the off-chance that they are essential to survival. Humans, as the pinnacle of natural selection on Earth, are pattern-seeking individuals, but the patterns we find are biased.

Looking for reason

Do the lines look straight to you? Or is your brain searching for a pattern that does not exist?

Tihs platren-skenieg aibilly can be swhon in a nmebur of dfeferint wyas. Aocdcrnig to rseecrah, it dseno't mttaer in waht oderr the lterets in a wrod are, the huamn mnid can sltil raed it. Tihs is bucseae the huamn biran deos not raed ervey ltteer by istlef, but the wrod as a wlohe. The olny irpoamtnt tihng is taht the frsit and lsat ltteer are in the rhgit pclae; our barin flisl in the rset as it sekes a ptatren.

The bias of our brains can also be shown in optical illusions, where again our brains fit our observations to common sense. To perceive a true picture of the world, we must contrast and compare the patterns we see against those of others to remove any bias. This is the very heart of the scientific method.

This book

This book is the story of one of the greatest pattern-seeking accomplishments of humankind: the Periodic Table of Elements. To understand construction of the table we start with the lessons learned by European thinkers when reading writings from the ancient world, then look at the birth of chemical experimentation in the Dark Ages, when alchemists sought connections in nature. As elements continued to be discovered across the globe, and patterns formed, many tried to sort the elements according to various criteria. Next we consider the stroke of genius of Dmitri Mendeleev and how his work differed from that of previous scholars.

To understand the behaviour of each element we then dive inside the atom from which they are made. Discovery of atomic structure, and eventually the modern quantum atom, provides fundamental understanding of an element's behaviour and placing in the table.

The rest of the book is dedicated to the individual elements: uses arising from their behaviour and tales of their discovery. After all 118 stories have been covered we end by discussing what the future holds for the table and the possibility of more elements to come.

Constructed from centuries of comparison of many different insights into nature, the Periodic Table is a testament to the scientific method but also our evolved ability to recognise bears in bushes.

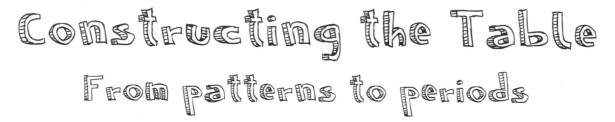

Constructing the Table
From patterns to periods

Revolutions in thinking about the world around us swept through Europe in the first half of the 17th century. Ancient texts from Greece and Rome, once thought lost, were found in the libraries of the Arabic world. A new generation of European thinkers were privileged to share in the thoughts of Aristotle, Plato and many others on the matter of natural philosophy. Driven by the ease at which these texts could now be produced by movable-type printers, this was an age of rediscovery and renaissance and the beginning of the scientific revolution.

From alchemy to science

Natural philosophers of the time linked scientific thinking as we see it today with theology (religions) and metaphysics (the idea of being). For various reasons they were looking for connections, physical or spiritual, embedded in the world; some practised 'magia', a precursor to science, desiring to learn of those connections so that they might use them for some practical end. Alchemists were one such group: they had existed since the Middle Ages, and their aim was not only to find these connections, but to purify and perfect objects.

One practical aim that many of the groups had was to find an object that transformed common metals such as lead and mercury into precious gold. Discovery of this 'Philosopher's Stone' was the goal of Hennig Brand, who spent all of his own money and the money of two wives searching for this mythical object. Although this was an ancient quest, Brand decided to use very modern methods of investigation. He experimented with human urine, and through heating and distilling, and then mixing the resulting residues, he found himself left with a glowing white substance. Without knowing, he had become the first person to chemically discover a new element; he named the glowing white substance phosphorus.

Engraving of 1771 artwork by Joseph Wright of Derby, thought to show German alchemist Hennig Brand discovering phosphorus in 1669.

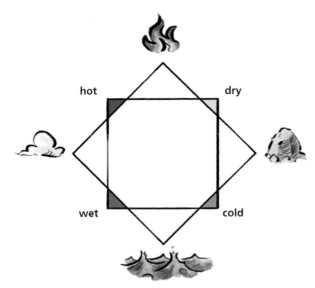

The four ancient elements of earth, air, fire and water which Aristotle believed made up everything in the world around us. The elements were related by four qualities: hot, cold, dry and wet.

Modern chemistry and elements

As the years passed, the scientific revolution gathered pace. Many substances had been identified as unique in property and now scientists began to compare each of them to the other. In 1661 the Irish-born natural scientist Robert Boyle wrote what most consider the founding book of modern chemistry: *The Sceptical Cymist*. Boyle rejected the ideas of Aristotle that everything is made from the four elements of earth, air, fire and water. Instead Boyle expounded the modern idea that chemical elements were 'perfectly unmingled bodies…not being made of any other bodies', although he then continued by stating that there were not any known substances with 'perfectly unmingled bodies', not even gold, silver, lead, sulfur or carbon. As simple as the definition might seem, Boyle's idea stood for over two centuries until the discovery of subatomic particles (see Atomic Physics).

In the years that followed, chemical experimentation resulted in the identification of more seemingly elemental substances. Whatever scientific method was employed, it seemed that such substances could not be reduced or separated further. Careful observation showed that some of these elements behaved in similar ways in similar experiments, but in others their behaviour was vastly different. Being the pattern-seeking animals we are, many scientists decided to look for underlying reasons for the results seen.

Starting to get things in order

In 1789 French nobleman and all-around self-promoter Antoine-Laurent de Lavoisier wrote *Traité Élémentaire de Chimie* (*Elementary Treatise of Chemistry*). Lavoisier identifies a number of 'simple substances…which may be considered the elements of bodies' and goes further to classify them as metallic and non-metallic substances. (The words metal and metallic come from the Greek and then Roman words for a mine – *métallon*, *metallum* – because these substances were extracted from the earth by mining or quarrying.) This is the first published classification of elements into groups and they were categorised based on the results they displayed in certain chemical reactions.

In 1817 the German chemist Johann Wolfgang Döbereiner grouped a number of the known chemical elements into groups of three, which he called triads. These triads of chemical elements had related properties, and the atomic mass of the middle element was calculated to be the average of the other two. This model was solid in reasoning but disregarded a large number of other elements.

By 1860 there were some 60 known elements, within which French geologist Alexandre-Emile Béguyer de Chancourtois noticed a repeating pattern. He placed the elements upon a helix (a spiral coiling around a cylinder) in order of increasing atomic mass. The elements that showed similar properties seemed to line up underneath and above one another. This repeating periodicity of properties was an amazing discovery but de Chancourtois' realisation went largely unnoticed by

Predicted v actual atomic mass of the central atom of each triad

Element 1 Atomic mass	Element 2 Actual atomic mass Mean of 1 & 3	Element 3 Atomic mass
Lithium 6.9	Sodium 23.0 23.0	Potassium 39.1
Calcium 40.1	Strontium 87.6 88.7	Barium 137.3
Chlorine 35.5	Bromine 79.9 81.2	Iodine 126.9
Sulfur 32.1	Selenium 79.0 79.9	Tellurium 127.6
Carbon 12.0	Nitrogen 14.0 14.0	Oxygen 16.0
Iron 55.8	Cobalt 58.9 57.3	Nickel 58.7

This table shows the triad groupings of elements by Johann Wolfgang Dobereiner, which he used to predict the atomic weight of central elements as an average of the other two. The upper number in the central column is the prediction and below are the very similar measured values.

chemists. Having used geological and not chemical terms in his 1862 paper, and originally publishing without a diagram of this brilliant idea, his genius was not truly realised until after his model was surpassed by that of Dmitri Mendeleev, some seven years later.

Musical chemistry

With de Chancourtois in obscurity, the oblivious English chemist John Newlands was working on a classification method of his own. Like de Chancourtois, Newlands also noticed a periodicity in the properties of the elements, stating that 'the eighth element starting from a given one is a kind of repetition of the first, like the eighth note of an octave in music'. Classifying all 62 of the then known elements, in 1864 he was the first to use the term 'periodicity' for the observed repeating pattern of chemical properties. Newlands was also the first, in 1864, to assign each element an atomic number, which he used to accentuate his Law of Octaves, a name he coined a year later. Most powerful of all was the new classification system's ability to make predictions; an essential part of any scientific model. Gaps that Newlands

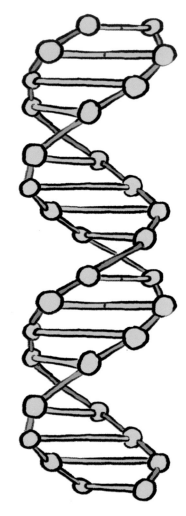

French geologist Alexandre-Emile Béguyer de Chancourtois' 1862 helical arrangement of the elements.

The secret life of the periodic table

English Chemist John Newlands' Law of Octaves likened the repeating pattern of the elements to musical octaves.

left in his table suggested that there might be as yet 'unknown, or, perhaps, in some cases only unrecognised' elements. While most of his suggestions of this type were later dismissed as incorrect, he did predict the 'at present wanting' of an element between silicon and tin, later discovered in 1886 (see Germanium).

As with so many discoveries ahead of their time, Newlands' idea was ridiculed by his peers, so much so that a lecture he presented was not published as usual by the Chemical Society. A reason for the denial of publication could well have been the nefarious intentions of the then Secretary of the Chemical Society, William Odling, who was also working on a scheme for classifying the elements. It was not until 1887 that the Chemical Society recognised Newlands' work; in 2008 they honoured his contribution as the 'discoverer of the Periodic Law for the chemical elements' with an inscribed blue plaque hung on the house in which he was born.

Odling's work, which he published the same year as Newlands, was also in its way ahead of the time. He sorted the elements into repeating units of seven. He correctly identified that iodine should take place in the group after thallium despite it having a smaller atomic weight, something that Dmitry Mendeleev didn't get right on his first attempt. He was also able to correctly group lead, mercury and platinum, something that his contemporaries missed. Odling was not given recognition because of his pivotal role in discrediting Newlands' work.

All of this work by English, French and German scientists laid the foundations for our modern periodic table of the elements. Without the keen eyes and pattern-seeking abilities of these scientists, Dmitri Mendeleev would not have been able to formalise all of these ideas into the table we have today.

Mendeleev and the Modern Table

Dreaming of modern chemistry

Born in the bleakness of Siberia, Dmitri Mendeleev was the youngest of a long list of siblings (sources vary, but he is understood to have had either 11, 13, 14 or 17 brothers and sisters!). After losing his father at the age of 13 and following the loss of the family business in a fire, a young Dmitri was moved around Russia by his mother in search of a higher education. After refusal from the university in Moscow he was given a place in St Petersburg, at the university his father had attended, and was followed to the city by the remaining and now poor Mendeleev family.

After completing his studies, Mendeleev contracted tuberculosis and so moved to Crimea, an area long praised for the healing power of its waters, where he took a post as a science teacher. Returning to St Petersburg in 1857 with a clean bill of health, he married, gained his doctorate, and gained tenure over the next 10 years

Dreaming of elements

Now teaching at the university, Mendeleev wrote the definitive chemistry textbook of his time: *Principles of Chemistry* (two volumes, 1868–70). While writing the textbook he was said to have envisaged the periodic table in a dream: 'I saw in a dream a table where all elements fell into place as required. Awakening, I immediately wrote it down on a piece of paper; only in one place did a correction later seem necessary.' Whether this was the truth or poetic licence after the fact, it was writing the book that forced Mendeleev's attempt to classify the elements according to their chemical properties. In 1869 he presented to the Russian Chemical Society his idea of ordering and classifying the elements.

Dmitri Mendeleev, father of the modern periodic table of elements.

ОПЫТЪ СИСТЕМЫ ЭЛЕМЕНТОВЪ.

ОСНОВАННОЙ НА ИХЪ АТОМНОМЪ ВѢСѢ И ХИМИЧЕСКОМЪ СХОДСТВѢ.

```
                              Ti = 50    Zr = 90     ? = 180.
                              V = 51     Nb = 94    Ta = 182.
                              Cr = 52    Mo = 96    W = 186.
                              Mn = 55    Rh = 104,4  Pt = 197,4.
                              Fe = 56    Rn = 104,4  Ir = 198.
                           Ni = Co = 59  Pl = 106,6  O- = 199.
         H = 1                Cu = 63,4  Ag = 108   Hg = 200.
              Be = 9,4 Mg = 24  Zn = 65,2  Cd = 112
              B = 11   Al = 27,4  ? = 68   Ur = 116   Au = 197?
              C = 12   Si = 28   ? = 70    Sn = 118
              N = 14   P = 31   As = 75    Sb = 122   Bi = 210?
              O = 16   S = 32   Se = 79,4  Te = 128?
              F = 19   Cl = 35,6 Br = 80   I = 127
         Li = 7 Na = 23  K = 39  Rb = 85,4  Cs = 133   Tl = 204.
                        Ca = 40  Sr = 87,6  Ba = 137   Pb = 207.
                        ? = 45   Ce = 92
                     ?Er = 56   La = 94
                     ?Yt = 60   Di = 95
                     ?In = 75,6 Th = 118?
```

Д. Менделѣевъ

Mendeleev's original table of elements published
in 1869 which shows the beginnings of the modern
periodic table.

Without knowledge of work done by his
English, French and German contemporaries
Mendeleev not only summarised all of their
work but extended their ideas. He first noted the
periodic repetition of properties when the elements
are arranged in order of their atomic mass, but it
was the immense predictive power of Mendeleev's
table that made it a truly fantastic scientific model.
Not only did the table have the ability to predict
the existence of new elements, but also how

they might be found. The patterns predicted how elements react with other chemicals which was the key to unlocking their discovery.

Patterns of behaviour

Mendeleev remarked that elements side by side with similar atomic mass exhibited a similar degree of reactivity with other chemicals; these constituted the rows in the table, which he named periods. Also mentioned was a similarity in chemicals produced in reaction of elements where the atomic mass increments regularly; these were aligned in columns of the table called groups. Mendeleev's design highlighted these patterns, which laid down the groundwork for the periods and groups of the modern periodic table.

Another trend that Mendeleev spotted in the table was British chemist Edward Frankland's idea of the 'combining power' of the elements, which today we call valency. Frankland noted in 1852 that different elements had desires to form compounds containing a certain number of additional atoms. He noted that 'nitrogen, phosphorus, antimony, and arsenic especially exhibit the tendency of these elements to form compounds containing 3 or 5 equiv[alence]s of other elements'. Mendeleev saw that the atomic mass ordering of the elements echoed that of valencies. He commented that this can be seen most clearly in the series: Lithium (1), Beryllium (2), Boron (3), Carbon (4), Nitrogen (5), where the number in brackets shows the maximum valence of the element (number of other atoms it bonds to in forming compounds).

In 1864, the German chemist Lothar Meyer published a book (unbeknown to Mendeleev) in which he arranged 28 elements into six families by order of their valence. Meyer's model demonstrated the periodicity of valence but he stopped short of making any predictions as to the existence or properties of undiscovered elements. We now know that valency is determined by the number of electrons participating in chemical reactions, which we call valence electrons. Mendeleev sent his 1869 paper to all of the

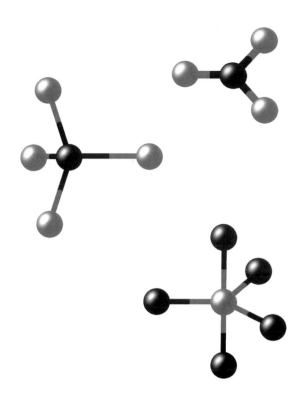

The valence of an element is a measure of how many bonds it can form with other elements. This diagram shows, from left to right, central atoms with a valence of 3, 4 and 5.

eminent chemists of the time and Meyer was on that list. When he received the paper and noticed the pattern of valency in the table, Meyer published an expanded and updated version of his 1864 work which closely resembled that of Mendeleev. Both Meyer and Mendeleev were recognised for their contribution in classifying the elements in 1882 when they both received the Davy Medal from the Royal Society.

Predicting the unseen

Like Newlands, Mendeleev's tabling of the elements included gaps where otherwise observed patterns were not seen to repeat. New, as yet undiscovered, elements were suggested to reside within these gaps and from the patterns in the table a prediction could be made as to the properties of each. Originally Mendeleev predicted the existence of four elements which he named eka-boron, eka-aluminium, eka-manganese and eka-silicon. The predictions regarding the properties of these elements agree well with the eventual discovery of scandium, gallium, technetium and germanium.

The eka- prefix, along with the dvi- and tri- prefixes Mendeleev used in his later tables, are ancient Indian Sanskrit for the digits 1, 2 and 3. Mendeleev used them to denote the undiscovered element being 1, 2 or 3 places below the named known element in his table; for example, eka-aluminium occupies the gap one period directly below aluminium in his table. The choice of Sanskrit is most likely a dedication to the ancient Indian scholars who developed the language. While the grammarians based Sanskrit on a two-dimensional pattern of basic sound made by our mouths, Mendeleev constructed his table from a two-dimensional pattern of repeating chemical properties.

Mendeleev also expressed his concern that the atomic mass of certain elements as understood at the time were wrong. The atomic weight of tellurium, he said, could not be 128 as measured at the time, but instead must lie, according to his table, between 123 and 126. Although correct about most concerns, Mendeleev was wrong in this instance (see Tellurium).

Elusive elements

Hydrogen seemed not to have a place in the table as it exhibited behaviour seen in elements from various groups. For this reason it was simply placed at the top above group 1. The table's predictive power was great but it could not have predicted the existence of an entire group of elements: the noble gases. The reluctance of these elements to chemically react meant that they were not observed in chemical reactions and so could not be isolated using techniques at the time. It was not until the advent of liquefaction of air and atomic identification through spectroscopy that these aloof gases were seen for the first time.

The Swedish giant of chemistry Jöns Jacob Berzelius was the father of the chemical symbol. He began using shorthand to record his many experiments, also introducing the number after the symbols to denote the number of atoms of each element present in a compound. Although Berzelius used a superscript number we today use a subscript: the familiar molecule of water, with two hydrogen and one oxygen, would have been written by Berzelius as H^2O, while today we write it as H_2O to prevent confusion with mathematical equations.

Periodic Table

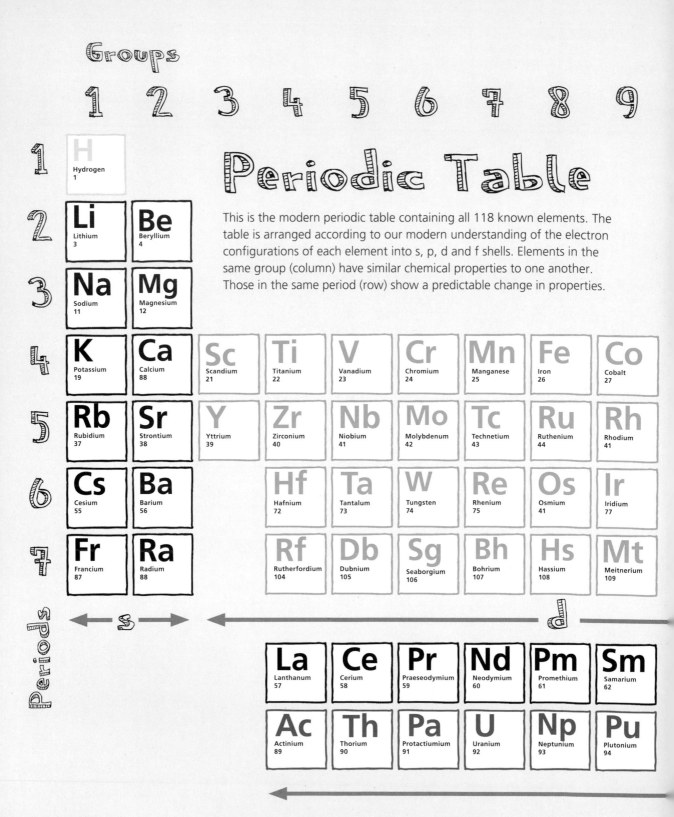

Groups

1 2 3 4 5 6 7 8 9

Periods

	1	2	3	4	5	6	7	8	9
1	H Hydrogen 1								
2	Li Lithium 3	Be Beryllium 4							
3	Na Sodium 11	Mg Magnesium 12							
4	K Potassium 19	Ca Calcium 88	Sc Scandium 21	Ti Titanium 22	V Vanadium 23	Cr Chromium 24	Mn Manganese 25	Fe Iron 26	Co Cobalt 27
5	Rb Rubidium 37	Sr Strontium 38	Y Yttrium 39	Zr Zirconium 40	Nb Niobium 41	Mo Molybdenum 42	Tc Technetium 43	Ru Ruthenium 44	Rh Rhodium 41
6	Cs Cesium 55	Ba Barium 56	Hf Hafnium 72	Ta Tantalum 73	W Tungsten 74	Re Rhenium 75	Os Osmium 41	Ir Iridium 77	
7	Fr Francium 87	Ra Radium 88	Rf Rutherfordium 104	Db Dubnium 105	Sg Seaborgium 106	Bh Bohrium 107	Hs Hassium 108	Mt Meitnerium 109	

This is the modern periodic table containing all 118 known elements. The table is arranged according to our modern understanding of the electron configurations of each element into s, p, d and f shells. Elements in the same group (column) have similar chemical properties to one another. Those in the same period (row) show a predictable change in properties.

s ← →   ←   d

La Lanthanum 57	Ce Cerium 58	Pr Praeseodymium 59	Nd Neodymium 60	Pm Promethium 61	Sm Samarium 62
Ac Actinium 89	Th Thorium 90	Pa Protactiumium 91	U Uranium 92	Np Neptunium 93	Pu Plutonium 94

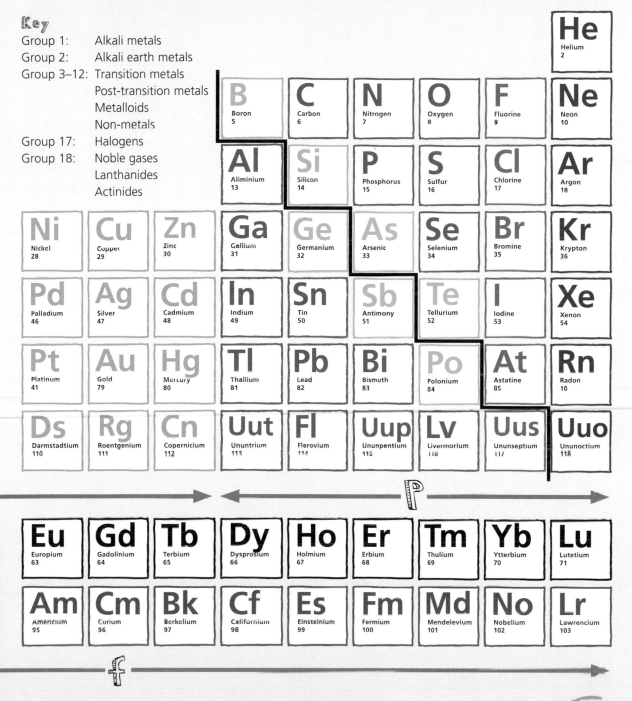

10 11 12 13 14 15 16 17 18

Key
Group 1: Alkali metals
Group 2: Alkali earth metals
Group 3–12: Transition metals
Post-transition metals
Metalloids
Non-metals
Group 17: Halogens
Group 18: Noble gases
Lanthanides
Actinides

He Helium 2

B Boron 5 | C Carbon 6 | N Nitrogen 7 | O Oxygen 8 | F Fluorine 9 | Ne Neon 10

Al Aliminium 13 | Si Silicon 14 | P Phosphorus 15 | S Sulfur 16 | Cl Chlorine 17 | Ar Argon 18

Ni Nickel 28 | Cu Copper 29 | Zn Zinc 30 | Ga Gallium 31 | Ge Germanium 32 | As Arsenic 33 | Se Selenium 34 | Br Bromine 35 | Kr Krypton 36

Pd Palladium 46 | Ag Silver 47 | Cd Cadmium 48 | In Indium 49 | Sn Tin 50 | Sb Antimony 51 | Te Tellurium 52 | I Iodine 53 | Xe Xenon 54

Pt Platinum 41 | Au Gold 79 | Hg Mercury 80 | Tl Thallium 81 | Pb Lead 82 | Bi Bismuth 83 | Po Polonium 84 | At Astatine 85 | Rn Radon 10

Ds Darmstadtium 110 | Rg Roentgenium 111 | Cn Copernicium 112 | Uut Ununtrium 113 | Fl Flerovium 114 | Uup Ununpentium 115 | Lv Livermorium 116 | Uus Ununseptium 117 | Uuo Ununoctium 118

P

Eu Europium 63 | Gd Gadolinium 64 | Tb Terbium 65 | Dy Dysprosium 66 | Ho Holmium 67 | Er Erbium 68 | Tm Thulium 69 | Yb Ytterbium 70 | Lu Lutetium 71

Am Americium 95 | Cm Curium 96 | Bk Berkelium 97 | Cf Californium 98 | Es Einsteinium 99 | Fm Fermium 100 | Md Mendelevium 101 | No Nobelium 102 | Lr Lawrencium 103

f

Atomic Physics
The smallest parts of an element

Atoms

The ancient Greek philosophy of atomism suggests that if we can understand the smallest building blocks from which a thing is constructed then we will truly understand that object. This idea was adopted by natural philosophers, who are today's modern scientists, in a search for the smallest atomic units of nature: atoms, derived from the Greek *atomos*, meaning uncuttable.

Chemistry of the 18th century had shown that some chemicals were compound combinations of other simpler chemicals. The first years of the next century saw many chemists engaged in carefully measuring the 'combining weights' of these compound chemicals. English chemist John Dalton demonstrated that you can infer the relative weight of the simple parts which react to form a compound chemical. Dalton's atomic theory showed that chemicals react in discrete whole number combinations, and he provided a table in his paper of the weights of these simple units relative to that of hydrogen.

For over a century many scientists remained sceptical about the existence of such chemical atoms. That was until 1905, when a previously unknown Swiss-based patent clerk – one Albert Einstein – used them to explain the bizarre phenomenon of Brownian motion. Peering into his microscope in 1827, the botanist Robert Brown had noticed that dust particles in water moved about erratically. Einstein explained that this random movement could be mathematically described if dust were colliding with discrete, atomic, units. French physicist Jean Perrin used the theory to determine the size and mass of these tiny atoms through an experiment in 1908.

Inside the atom

Even before this atomic theory received Einstein's long-awaited confirmation it was already being superseded. Richard Laming, a surgeon by day but a scientist by night, was regarded by the scientific establishment as eccentric. He published a number

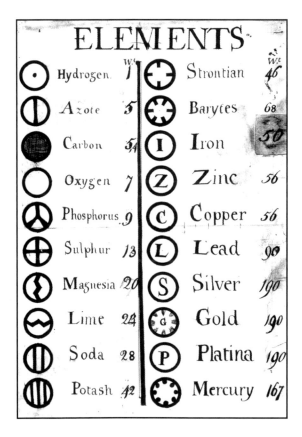

John Dalton's 1808 table of atomic weights and the symbols he used for a number of 'elements'. Some of the substances he includes are now known to have been compounds, made of two or more true elements.

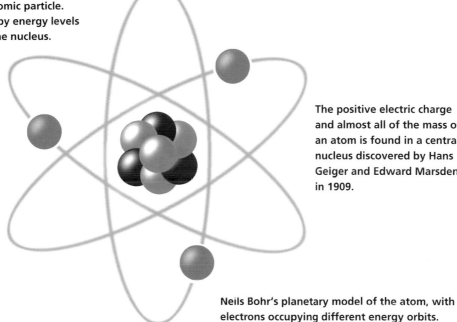

In 1897 in Cambridge, UK, J.J. Thomson discovered the electron, the first subatomic particle. Electrons occupy energy levels surrounding the nucleus.

The positive electric charge and almost all of the mass of an atom is found in a central nucleus discovered by Hans Geiger and Edward Marsden in 1909.

Neils Bohr's planetary model of the atom, with electrons occupying different energy orbits.

of papers between 1838 and 1851 with ideas about a basic unit of charge which is responsible for the chemistry of the elements. Experiments in the late 19th century brought this idea into the mainstream as many scientists searched for this 'atom of electricity', given the name 'electron' by Irish physicist George Johnstone Stoney in 1891.

Stoney and others paved the path for J.J. Thomson and colleagues who, in Cambridge, England in 1897, made the crucial measurement. Thomson was experimenting with cathode ray radiation emitted from high electrically charged metal plates. He saw that their paths changed course in the presence of a magnet. This result showed that cathode rays were not like other radiation but instead made from electrically charged particles which had a mass much lighter than any measured chemical atom.

With the discovery of the electron came ideas of how an atom which contained electrons might look. Atoms were known to be electrically

neutral because, unlike the electron, the paths they travelled along were not deflected by a magnet. Thomson imagined his negatively charged electron particles distributed evenly in a sea of positive electric charge within the atom. This 'plum pudding' model of the atom, as it came to be known, was put to the test in Manchester, England by researcher Hans Geiger and his student Ernest Marsden, under the watchful eye of the head of physics at the university, Ernest Rutherford. They used Rutherford's recently discovered alpha particle radiation (see Helium) as probes to look inside larger atoms of gold. When they bombarded a thin foil of gold with the positive electrically charged particles, most breezed right through. On rare occasions, though, an alpha particle was seen bouncing back from the foil like a ball against a wall.

This observation demonstrated that the positive charge in an atom could not be evenly distributed but instead was located all in one small region. Only a very highly concentrated positive electric charge could deflect the energetic, and also positively charged, alpha particles. This led Rutherford to imagine an atom where Thomson's negatively charged electrons orbited like planets around this densely packed positive atomic nucleus. We know today that the positive electric charge in the nucleus comes from smaller particles called protons. Alongside the protons in the nucleus there are also electrically neutral neutrons (see Iron). Such a planetary atom would be unstable though, because electrons in planetary-like orbit would quickly spiral inwards, strongly attracted by the opposite electric charge of the nucleus.

Structure of the atom

Throughout the 19th century many chemical elements were discovered not through chemical reaction, but by looking at the light they emitted. Elements do not release a continuous spectrum of light, only certain specific colours. These form striated spectral line patterns which provide a unique, barcode-like, fingerprint identifying an atom. When a new spectral line was observed it had to come from a new chemical element. Our perception of the colour of light arises from its energy, and in the 1880s Johannes Rydberg linked the spectral lines to some, unknown, arrangement of energy levels within an atom. He demonstrated this by identifying a number of common spectral patterns among different atoms. At the turn of the 20th century Danish physicist Niels Bohr extended the idea to connect these spectral line energies of light emitted to the energy of electron orbits within an atom.

Just as energy is required to lift something to a height off of the ground, an electron would require energy to be raised higher to an orbit further from the nucleus. An object raised above the ground will have gained gravitational energy. This is paid back in kinetic (movement) energy if we drop it. A negative electrically charged electron raised to an orbit further from a positive nucleus

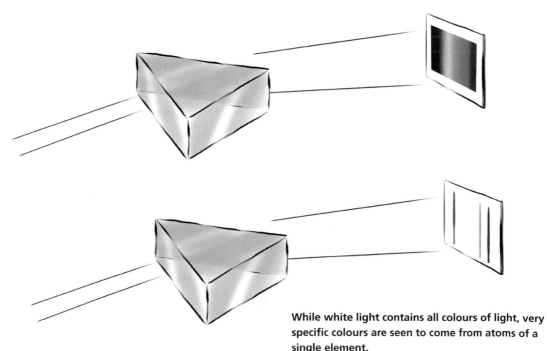

While white light contains all colours of light, very specific colours are seen to come from atoms of a single element.

The secret life of the periodic table

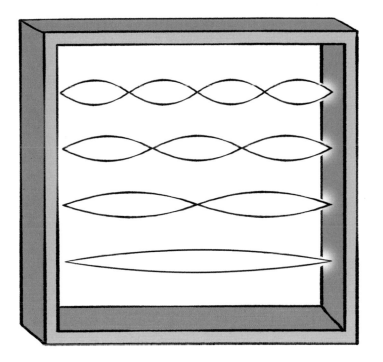

The higher the energy of a vibrating string the more complex the vibration becomes, demonstrating more areas of maximum and minimum vibration.

gains electrical energy. Electrical energy is gained or paid back by absorbing or emitting light. The energy of the light absorbed or emitted depends upon the energy difference between electron orbitals in the atom. If light coming from atoms can only have certain energies, Bohr explained, then this shows that electrons can only orbit with certain energies within atoms.

Discrete energies like these are also seen in the vibration of strings that are fixed at each end, like those on a guitar. If we pluck with little energy we may only produce maximum vibration at one central point on the string and two points of zero vibration at each end where the string is attached. If we strike the string harder we could produce an additional point of vibration, at two points on the string. Continue to increase the energy we give to the string and we continue to increase the number of these modes of vibration, beginning to notice a pattern. Each next higher energy vibration mode simply adds one point of vibration, and one stationary point. The energy of each string is related directly to this initial lowest mode vibration. Each wave is a whole number multiple, n, of this

smallest quantity. Because of this we say that these waves are quantised, made from some multiple of a quantum number n.

Vibrating strings like this are a good model for Bohr's electron orbitals in an atom. Each orbital in his model that is further from the nucleus is just an increase in a similar single quantum number.

Going back to our gravitational analogy, if we were to raise and then drop an object to the floor, its overall change in gravitational energy would be zero. If an electron were to spiral inward into the nucleus of an atom then, in the same way, its energy would also fall to zero. Bohr showed, however, that electron orbitals are quantised, like vibrations of a string, and can only exist if they have an energy that is some multiple of some quantum. This means that for an electron to orbit an atom it cannot have an energy which is lower than that of the smallest quantum of energy. Therefore, the orbital energy can never become zero and so electrons cannot spiral forever in toward the nucleus. This is why atoms are stable and give the first glimpse into the strange new world of the quantum atom.

The Quantum Atom
The basis of chemical behaviour

The bizarre behaviour of subatomic particles gave birth to the field of quantum physics. This field of science today gives us our most detailed picture of the atom. Arrangement of elements and the properties they exhibit are shown to arise naturally from this fundamental model.

Light and matter

In 1801 Thomas Young overthrew Isaac Newton's idea that light is made up of particle-like lumps which he called corpuscles. When shining light through two narrow slits Young showed that a series of light and dark lines are seen projected onto a screen. The pattern could only be explained if light were acting like a wave. Waves spread outward from each slit like ripples in a pond. As the ripples cross one another the peak of one may come into contact with the peak of the other. Constructive interference like this would lead to summation of the peaks, forming a wave with a higher peak. Alternatively, the same peak might meet the bottommost trough of the other wave.

This would result in a destructive interference as the peak completely fills the trough of the other, cancelling out the wave all together. It is this interference that results in the constructive bright fringes and deconstructive dark fringes in Young's experiment: light was a wave!

Just over 100 years later the tables seemed to turn as once more, thanks to another of his amazing 1905 papers, Einstein gave an answer to a very puzzling observation. When violet light is shone onto a piece of metal, electrons are liberated from the surface. Increasing the energy of a water wave would wash more sand from a beach; increasing the energy of the light, however, did not increase the number of electrons liberated

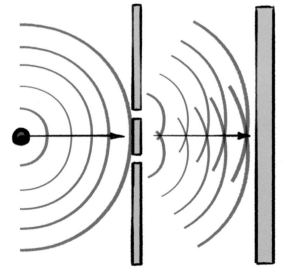

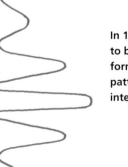

In 1801 light was seen to behave like waves, forming an interference pattern as one wave interacts with another.

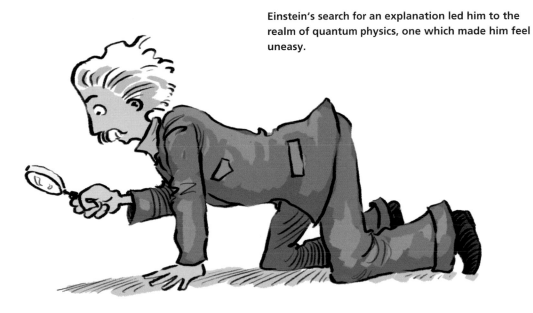

Einstein's search for an explanation led him to the realm of quantum physics, one which made him feel uneasy.

from the metal. It turned out that the only way to liberate more electrons was to make the light brighter, throwing more light at the surface of the metal. Einstein explained mathematically that this can only happen if light were acting like localised lumps of energy; much like Newton's particles.

Quanta

Light, it seems, is neither a classical particle nor wave but something new altogether. Einstein showed that light behaves like a lumpy particle when it is interacting with electrons. Young's interference demonstrated that when not interacting it interferes like a wave. Light is made up of quanta: packets of energy which we call photons.

The energy of the light in Young's double slit experiment did not change. This would have been seen as a change in colour. The light did not become lower in energy and more red where the waves deconstructively interfered, nor did it gain in energy, becoming more blue, where light combined constructively. Instead the light retained its colour and simply became brighter or dimmer on different areas of the screen. This was a change in the number of photons arriving at each

location on the screen, not the energy of each photon. This wave-like interference of light determines, then, the probability that photons find themselves projected onto a certain part of the screen.

The paths photons take through the slits to the screen are subject to probability, which Einstein summarised as the roll of a dice. This means you can never be certain where a photon will end its journey, even if you know how that journey began. Instead you can only calculate the probability that a photon will exist in different locations. This is very different to the determinate laws of physics up to this point, with which one can calculate a unique final outcome from an initial set of conditions. Einstein's paper was the spark that ignited the field of quantum physics, but its indeterminate nature worried him until his final days.

Louis de Broglie in 1923 suggested that electrons, protons, neutrons and atoms also behave in the same strange way as light. When observed they would behave like lumpy localised particles, but all other times they act like a probability wave. Experimental proof of this theory came in 1927 when George Paget Thomson in England, and Clinton Davisson and Lester Germer in the US saw Young's interference patterns using a beam of electrons.

Erwin Schrödinger, Werner Heisenberg, Max Born and Pascual Jordan each independently developed a theory using de Broglie's idea of 'matter waves'. The result was a mathematical tool to calculate the probability of a subatomic particle being found at a certain location, with a given energy. As it is concerned with the mechanical movement of quantum things, this theory became known as quantum mechanics. The humble hydrogen atom played an important role in the validation of quantum mechanics when it was used to predicted the structure of electrons within a hydrogen atom (see Hydrogen).

Waves in three dimensions

While Bohr's strings described earlier are a fine analogy of electron energies in the atom, they are not the whole picture. We know now that the vibration strength of a quantum wave represents not energy but the probability of finding an electron in a particular location; the probability is in truth the size of the vibration at a location (amplitude) squared (multiplied by itself). The energy of a quantum wave is represented by the number of points at which the wave is vibrating. As discussed, this is akin to guitar strings in a box and the energy of each string can be defined by

one quantum number, n.

In the lowest energy string, we would predominantly expect to find the electron in the centre of the box. As we continue increasing energy there are more and more positions in which the electron is equally likely to be found. But you can only go back and forth along a string. This two-way motion represents just one dimension of space. We live in three spatial dimensions; up and down, back and forth, and left and right. Modelling the energy of electron orbitals in three-dimensional atoms requires you to move from one-dimensional strings to three-dimensional (3D) clouds. These electron clouds trace the most likely location to find electrons orbiting around a 3D atomic nucleus. As with strings, more modes of vibration are generated within the cloud with each increase in energy. This is seen as greater fragmentation in the shape of the clouds; they break up into a number of regions equally probable to contain electrons.

While the energy of a 1D string can be uniquely defined by a single quantum number, three quantum numbers are required to uniquely define 3D vibrational modes. One number, n, determines the maximum distance a cloud forms away from the nucleus and is defined by chemists as an electron shell. The second number, l, defines the number of vibrational modes and therefore

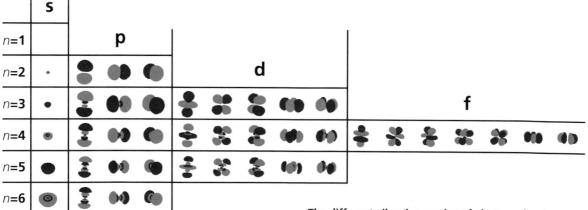

The different vibration modes of electron clouds, becoming ever more complex in shape as the electron orbit increases in energy.

the shape of each electron shell, defined by chemists as electron subshells. The third number, m, determines the orientation of each lobe of an electron cloud around the nucleus. As with the case of the string, the ground state lowest energy has value n=1 and l=0, m=0. The three quantum numbers are closely linked to one another; for any given value of n, l can only take values between 0 and n-1; m may take any value from –l to +l. This means that when n=1 the cloud is a sphere as there is zero fragmentation (as l=0) and no lobes (as m=0). The lowest energy an electron can have when n=2 is in the cloud n=2, l=0, m=0, which is just a higher-energy version of the ground state. With n=2 we can also have the subshell n=2, l=1 which contains three lobes as m can equal -1, 0, or 1. Each increase in n also creates a new cloud which has an increased number of lobes.

Whilst this numbering system is used in quantum physics, chemists label the electron subshells a little differently. The electron shell, n, is labelled in the same way, with a number, but subshells, l, are given a letter instead. The letters relate to the historical naming of atomic spectral lines, relating to how they appeared; l=0=s (for sharp), l=1=p (for principal), l=2=d (for diffuse), l=3=f (for fine). Any possible higher modes of l would just follow the alphabet, so the next would be l=4=g, but all known elements to date have at most an f subshell. An electron in the shell defined by the quantum numbers n=2, l=1 would be said by chemists to be in shell 2p, and n=3, l=0 in shell 3s.

Uncertain spin

Each unique lobe, defined by the three quantum numbers, can accommodate two electrons. Wolfgang Pauli correctly stated that no two identical electrons, or any particles making up an atom, can occupy the same space with the same energy. So for two electrons to occupy the same cloud lobe there has to be a property which allows them to be uniquely identified. This

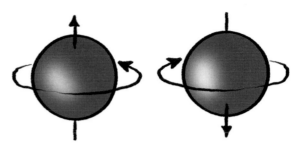

Electrons possess an internal 'spin' property similar to that of a spinning top which can be used to uniquely identify two electrons.

property is known as quantum spin and is similar in many ways to the spin of a spinning top. If spun clockwise onto a glass table, a spinning top would look from below to be spinning counterclockwise. This is very different to the same spinning top spun counterclockwise to begin with, which looks to be spinning clockwise from below. Just as each of these situations can be uniquely defined by eye, two unique internal spin-like states of electrons can be determined by experiment. This quantum spin introduces a new quantum number when identifying an individual electron taking the value of +1 or -1. This way two electrons can obey Pauli's exclusion principle and sit together inside a single lobe of an electron cloud.

With this in mind each s subshell (l=0; m=0) has just one lobe and can only accommodate two electrons. Each p subshell (l=1; m=-1, 0 or +1) has three lobes, each containing two electrons, making six in total. A d subshell (l=2; m=-2, -1, 0, 1 or 2) can contain 10 electrons and an f subshell (l=3; m=-3, -2, -1, 0, 1, 2 or 3) 14. Each subshell can contain 2(l+1) electrons.

Trends and Patterns
Following the signs and filling atoms

The periodic table is a powerful tool for predicting the way in which chemical elements will react, both with each other and with more complex compound chemicals. Chemical reactions occur when there is a sharing or exchange of an atom's outermost valence electron. It is an atom's ability to attract electrons to join their valence or their willingness to give them up which dictates the element's properties.

Atomic number and the size of atoms

The atomic number of an element defines the number of protons in the nucleus and in turn the number of electrons in orbit around them. As the atomic number increases, so too does the atomic weight of an element as it takes on more and

more mass. To ensure stability, neutrons are added to the nucleus and, dependent upon the number required, they add varying weight to each element. The atomic weight is calculated to be 1/12th mass of carbon-12, which itself contains 6 protons, 6 neutrons and 6 electrons. It is not a whole number because there are a number of factors which affect

6 protons

6 electrons

The smaller atomic number informs us of the number of protons within the nucleus of each atom. The larger atomic weight is an average of the atomic masses of the most stable isotopes of an element, which can contain additional or fewer neutrons alongside the protons.

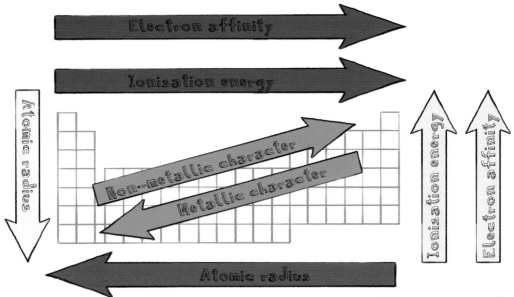

This outlines the general increase in atomic radius, ionisation energy and metallic character of elements in the periodic table.

the overall mass of an atom such as how tightly bound the protons are in the nucleus.

Groups are stacked into columns on the table and are collections of elements with similar chemical properties. The similar chemical properties arise from the fact that they have the same number of outermost, valence, electrons within the same type of subshell. Group 1 have just a single electron in their outer s subshell, group 2 have two in their outer s subshell.

As the atomic number increases, more electrons are piled into the atom. Just like a glass filling up, the outermost electrons added last become further and further away from the nucleus. If you calculate the radius of the atom it is seen therefore to increase with atomic number down a group. As atoms get larger it becomes easier for the outermost electrons to escape the lure of the ever-distant nucleus. The energy required to remove an electron from an atom, to make it a positive electrically charged ion, is known as the ionisation

energy. As the radius of an element increases down a group, the ionisation energy drops. But the trend does not work as we go across periods from left to right, and the reason for this is to do with the way electron clouds are filled.

How to fill an atom

The Aufbau principle, from the German for 'building up', was part of Bohr and Pauli's original concept for electron configuration inside the atom. The modern form retains much of this original idea and describes an ordering of electron subshell energies given by Madelung's, also known as Klechkowski's, rule. This rule was first stated by French engineer Charles Janet in 1929, rediscovered by German physicist Erwin Madelung in 1936, and later given a theoretical justification by Soviet chemist V.M. Klechkowski in 1962.

The rule states that shells are filled with electrons in order of increasing energy, with the lowest-energy shell filled before electrons are

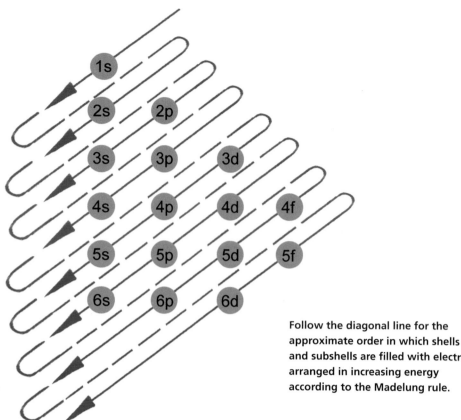

Follow the diagonal line for the approximate order in which shells and subshells are filled with electrons, arranged in increasing energy according to the Madelung rule.

placed in higher-energy shells. Both quantum numbers n and l affect the average distance from the nucleus, and therefore energy. The subshells are filled in order of increasing n+l. If the addition n+l is the same for two subshells then it is the shell with the lowest n that is filled first, e.g. between n=2, l=1 and n=3, l=0 it would be the n=2, l=1 subshell that is filled first. The diagram of the electron orbitals shows this leading to a diagonal rule for filling orbitals with electrons. One starts from the rightmost shell on each line and follows the diagonal red lines until the leftmost, l=0, column is reached. The filling then begins again from the next top rightmost shell.

The Aufbau principle works very well for the lightest 18 elements, then increasingly less well for the following 100 elements as other factors come into play. It is echoed in the layout of the periodic table. Groups 1 and 2 have their one- or two-

valence electrons in an s subshell. Groups 13–18 have their 1–6 outermost electrons in a p orbital, and the transition metals sandwiched between in a d orbital. The lanthanides and actinides which are shown as separate in the most popular form of table, as shown in this book, have their outer electrons in an f orbital.

I just want stability in my life!

All elements wish to exist in the lowest energy state possible and this means having the fullest and most symmetrical electron shell they can. Failing that, elements wish to have as complete and symmetrical an electron subshell as they possibly can. This is the main driving force for all of the trends in reactivity of the elements along a period (row of the table). As you go from left to right each member represents the addition of a proton to the nucleus and electron to the same

The secret life of the periodic table

electron shell. The noble gases at the end of each period in group 18 are not reactive as they have full electron shells and so do not need to exchange or share electrons to achieve stability. All other elements share or exchange electrons, in covalent or ionic chemical bonds, to achieve the same state. Those in the lower groups are happy to lose electrons as this brings them closer to the electron configuration of a lower atomic number noble gas. Those in the higher group, however, want to gain electrons as this is their quickest route to act like a noble gas. Sandwiched in between are the relaxed transition metals that are happy to give up their valence electrons in varying numbers.

Down a group, ionisation energy decreases as it becomes easier to lose electrons as they are evermore loosely held. But along periods it becomes increasingly more difficult to remove electrons as atoms want to keep them to complete their full electron shell. This means that on going from left to right along a period, there is a general increase in the ionisation energy. The top rightmost elements have higher ionisation energies and those of the bottom leftmost are low. Because of their desire to attract electrons rather than give them up, we also say that the top right elements on the table have a high electron affinity. The atomic radius also gets smaller along a period because

the nucleus of each element holds on tighter and tighter to their valence electrons, none more so than the noble gases of group 18.

Metal or not?

Broadly speaking the willingness to give up electrons is the trait of elements which we define as metallic. Those elements which desire electrons and are less willing to give them up are therefore non-metallic. Because of the diagonal trends mentioned above, a line can be drawn stepwise across the table to define metals and non-metals. The line, which can be seen on the table on pages 16 and 17, shows that the majority of elements are in fact what we would define as a metal, with a handful of non-metals. Those elements that find themselves on this dividing line show characteristics of both metals and non-metals under different circumstances. They are defined separately as metalloids (metal-like).

All of these trends are quite broad brush strokes over the finer details of the complex traits and characteristics of the elements. There are therefore many exceptions to the rule and these will be highlighted in the book. It was these trends, though, that led scientists from seeking patterns to uncovering the internal structure of the atom.

Large bottom of group 1 caesium is more than happy to let its outer (valance) electron roam free, while small top of group 7 fluorine atoms desperately hold tightly to theirs.

Trends Table

Of the many trends in the table the two which give the clearest picture as to the reactivity of an element are the size of the atom and its electronegativity.

As discussed in the last section, as atoms hold on tighter and tighter to their electrons the size of the atom decreases. Along a period the desire for an atom to keep its electrons increases as it gets closer and closer to a complete outer electron shell. Here we have filled the tile representing each element to different amounts to represent their relative size compared to the largest measured which is caesium. The less the tile is filled the smaller the radius of atoms of that element.

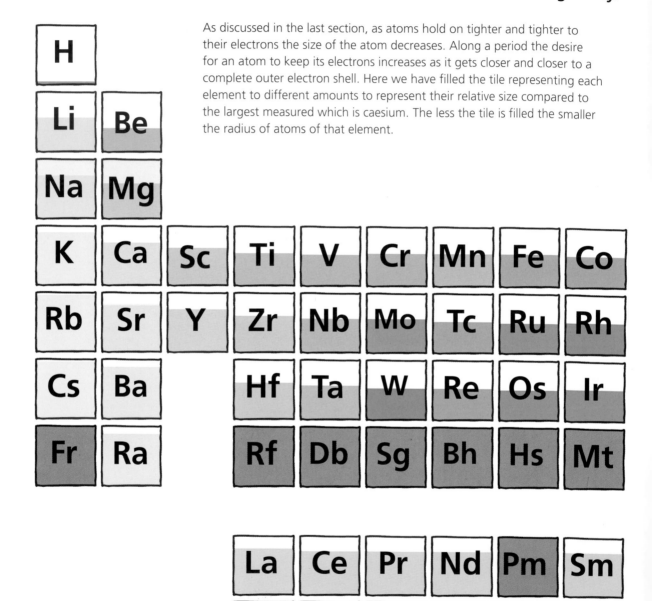

The secret life of the periodic table

The colour of the tile fill represents the electronegativity of that atom, scaled to the highest which is fluorine and the lowest which is caesium. As mentioned in the previous section this is a measure of how effective an atom of the element is at attracting electrons towards it. This shows the Pauling scale electronegativity, experimentally determined through investigation of the bonds that atoms form with different elements, named after American chemist Linus Pauling who first proposed the idea in 1932. For this reason it is not possible to obtain a value for some of the noble gases as they have not been observed reacting with more than one element or in some cases any element at all. The other greyed out boxes represent elements for which there is insufficient data.

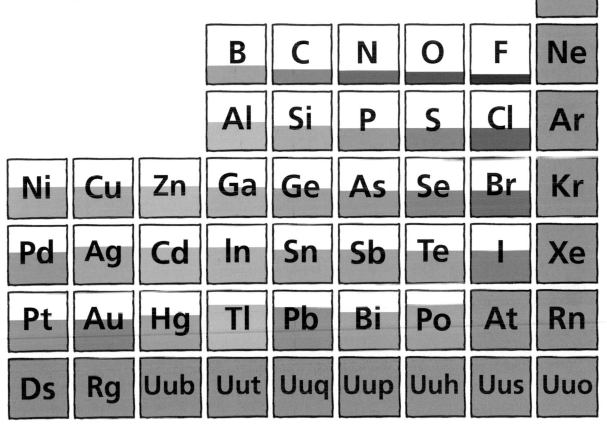

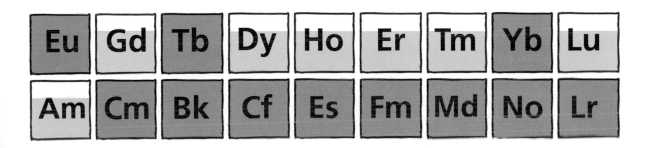

Hydrogen
Fiery and fundamental

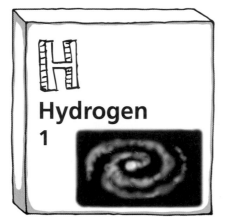

Hydrogen
1

Atomic number	1
Atomic weight	1.0082
Abundance	1400 mg/kg
Radius	25pm
Melting point	-259°C
Boiling point	-253°C
Configuration	1s1
Discovered	1766 – H. Cavendish

Hydrogen has come a long way in its almost 14 billion-year journey and it has an exciting future ahead of it.

Our tried and tested understanding of our universe stretches way back in time to within the first second after it was created. When just 10^{-35} seconds old, that's 0.000000000000000000 000000000000001, seconds old, the universe was a hot dense soup of subatomic particles. It took another 380,000 years of expansion and cooling before the first electrons could overcome the intense heat and become bound to protons or atomic nuclei, forming the first ever atoms. These first atoms were almost entirely the simplest atom possible with one proton and one electron: hydrogen. These primordial atoms were the only elements to exist for the next 100 million years.

The birth of a star
After 100 million years, hydrogen gas had formed clouds so large that they began to collapse under their own weight due to gravity. As the gas fell inward the cloud heated up rapidly in the centre. The increase in temperature and pressure was so high that particles were travelling at incredibly high

Under the right conditions protons can fuse together to form the heavier hydrogen isotope tritium. Tritium can then, under similar conditions, fuse to form a nucleus of helium – the second lightest element.

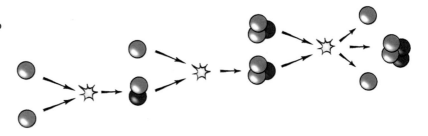

speeds, causing frequent and energetic collisions between the particles. These collisions allowed protons, neutrons and atomic nuclei to get close enough together that they fused to one another. This process is known as nuclear fusion and when it begins, a star is born. The fusion creates new heavier atomic nuclei from many lighter parts. All elements up to iron are created through this fusion process inside a star, but it is in the death of the largest stars that we get all of the heavier naturally occurring elements (see Iron).

Simple quantum physics

Hydrogen is still today the most abundant element, making up over 75% of the entire visible matter in the universe. The simple structure of the atom made it the perfect test candidate for quantum mechanics (see Quantum Atom). The two-body system of proton and electron in hydrogen is still the only atom for which the behaviour has been fully quantum-mechanically calculated, which demonstrated to many sceptics the power of this new science.

Does exactly as the name suggests

Although Robert Boyle noted a gas bubbling from a reaction between iron filings and dilute acids in 1761, it was Henry Cavendish who, five years later, recognised the gas as a unique substance. Cavendish originally called the gas 'flammable air' as it is extremely explosive; so much so that it was used by NASA as a fuel for the space shuttles. In 1781 Cavendish observed that when burnt, the new element produced water. In 1783 this prompted French chemist Antoine Lavoisier, after confirming Cavendish's discovery, to give the gas the name hydrogen, which almost literally translates from ancient Greek (*hydro-genes*) as 'water-creator'.

Nuclear past and future

An element is defined by its electron/proton number; adding or removing neutrons from the nucleus does not change an element. It is therefore

Inside the Joint European Torus nuclear fusion reactor at Culham, Oxfordshire, England, where isotopes of hydrogen are fused together to form helium.

possible to have a number of different weights of atom for each element – these are known as isotopes. There are two other naturally occurring isotopes of hydrogen; deuterium ^2H containing one neutron, and tritium ^3H containing two. Both have a role in nuclear energy. Deuterium is used in nuclear reactors as a neutron moderator (see Cadmium/Hafnium) in traditional fission reactors (see Uranium). Tritium is a possible nuclear fuel of the future: tritium fusion reactors are being developed to produce vast amounts of clean energy from nuclear fusion. It is the technical task of igniting and sustaining nuclear fusion outside of a massive star that presents the biggest challenge. Tritium is the best candidate as it requires the lowest amount of energy for fusion to occur, which should mean lower temperatures than inside a star.

Right now it requires more energy to initiate any fusion reaction on Earth than it generates, as high-power magnetic fields and lasers are required. In the future, though, this method of energy-generation could provide clean energy because the light elements made, like helium or lithium, are very safe.

Helium
The super element

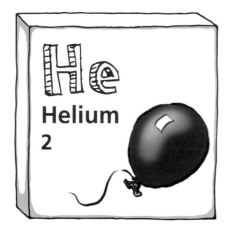

Atomic number	2
Atomic weight	4.0026
Abundance	0.008 mg/kg
Radius	no data
Melting point	-272°C
Boiling point	-269°C
Configuration	1s2
Discovered	1895 – Ramsay, Cleve & Langlet

Helium is the original ET; extra-terrestrial because it was the first element to be discovered outside of our atmosphere.

The solar eclipse in August of 1868 provided French astronomer Jules Janssen with a unique chance to observe the outer atmosphere of the sun. Janssen noted observing a yellow spectral line which he thought to be sodium; at the same time he discovered that the spectral line was bright enough to be observed even without an eclipse. In October of the same year English scientist Norman

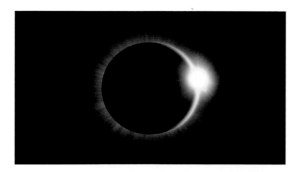

Helium gas was first seen in light coming from the outer atmosphere of the Sun.

Lockyer, despite the British weather, saw the same spectral line. Lockyer noticed that this line could not have come from sodium, lying instead in between two sodium spectral lines, and that it must be a new element. He and his colleague Edward Frankland later named this element 'helium' after the Greek sun god Helios. After realising the importance of his discovery, Janssen published his findings, and today both Janssen and Lockyer are credited for the element's discovery.

Back to Earth
It was almost another 30 years before terrestrial helium was discovered. In 1895 Swedish chemists Per Teodor Cleve and Nils Abraham Langlet discovered a gas naturally being released from rock containing the ore cleveite. The same year Scottish chemist William Ramsay also noticed a gas being released when reacting cleveite with acid. The Swedish pair were able to identify the atomic mass of the gas as a new element. Ramsey noticed that

the gas also had the same yellow spectral line as Lockyer's observations, which was confirmed by Lockyer himself.

Helium is extremely light and, just like ET, helium also wants to go home to outer space which means that although it is the second most plentiful element in the universe, it is rare to find it on our planet.

Nuclear birth

Ernest Rutherford showed in 1905 that alpha particles released during radioactive alpha decay were the nucleus of helium atoms. The helium released by cleveite was coming from the chain of radioactive decay associated with uranium. Build-up of helium gas from alpha particles is a real worry when burying radioactive waste from nuclear reactors. The longer radioactive waste is left, the more helium gas is produced from alpha particles. When burying radioactive waste for thousands of years you have to plan for the immense increase in pressure generated from all of this new gas, so the steel and concrete coffins the waste is buried in have to withstand huge pressures to avoid the nuclear waste exploding.

Super element

Helium has a complete set of valence electrons in the lowest energy 1s electron subshell which makes it the least reactive element of them all. Helium is a gas at room temperature as it is a collection of single atoms that really don't want anything to do with each other. It has a very low boiling point of just 4.2K (Kelvin, K, is measured in degrees above absolute zero), -268.9°C. Liquid helium is therefore very cold which makes it a great coolant. It is used to chill metals to the point where they become superconductors (see Yttrium) and have zero electrical resistance. Superconductors are used to create the world's most powerful magnets, which are used in everything from MRI medical scanning (see Niobium) to the world's highest energy particle accelerator, the Large Hadron Collider at CERN in Switzerland.

If you continue to cool helium then at around 2.17K, a temperature known as helium's lambda point, it begins to behave very strangely indeed, as it becomes a superfluid. Superfluids have zero viscosity (viscosity being physical resistance to flow). As a result of this, superfluid helium shows some very strange behaviour: for example, a cup of superfluid helium would gradually empty itself. All liquids are known to creep a little against gravity (take a look at the way water curves upward at the edge of a glass) and the force that drives this creep is something called a capillary force, which arises from interactions between the liquid and the solid container. The combination of gravity and viscosity does not allow liquids to creep far before this capillary force is defeated and the liquid falls back into the container. Zero viscosity superfluid helium, however, can creep up a surface unimpeded by viscosity and eventually spill over the edge of the cup.

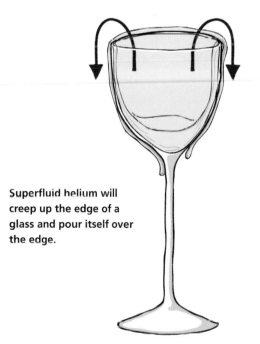

Superfluid helium will creep up the edge of a glass and pour itself over the edge.

Alkali Metals
Losing electrons and reacting raucously

Group 1 provides a fine example to demonstrate the amazing trends and patterns in the periodic table. Explosive and colourful, yet soft and essential for life, these alkali metals are also, at first glance, a group of contradictions.

All of group 1 are very reactive metals that must be stored in an inert, unreactive environment of mineral oil or noble gas. This precaution prevents them reacting with gases or water vapour in the air. Alkali metals are highly reactive because they have one electron in their outer s shell which they would desperately like to lose to make their outer shell complete, giving them an electron configuration like a noble gas. On reacting they tend to form positive ions (charged atoms) with a plus one charge after losing a negative electron $[0-(-1)=+1]$.

Afraid of being alone

The desire to react means that they are never found as a free element in the natural world and instead found bound to other elements in compounds. The compounds they are found in are mainly salts which are compounds containing metal(s) and non-metal(s) that are connected by an ionic bond; they have exchanged electrons between atoms and are consequently attracted to one another because of an equal and opposite electric charge.

The salts that alkali metals produce provide the name for the group. They all form alkaline

The vivid range of colours that the alkali metals show when excited by a flame. As the atomic radius increases down the group, so too does the reactivity and softness of these metals.

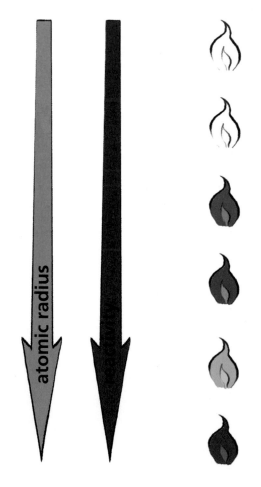

The explosive reaction of sodium with water shows sodium's distinct yellow light spectrum. The reactions of all alkali metals occur in the same way, producing hydrogen gas and a hydroxide solution, but increase with vigour as they become more reactive further down the group.

attached to the H+ acid ion. This is usually a non-metal and forms a new type of salt.

Putting on weight

As you go down the table the atoms get bigger as the number of electrons orbiting the nucleus increases; the addition of protons to the nucleus also contributes to giving each successive atom a greater mass and density. With an increase in size, the outer lone electron becomes further away from the central nucleus and the attraction to the nucleus is diminished. Because of a looser grip of the outer electron, the reactivity of the metals increases down the group: for example, lithium fizzes gently when placed in water while caesium explodes violently. Also going down the group the metals become softer and softer, because the less they wish to hold onto their outer electron the less they want to use it to form links between atoms of the same element. The melting and boiling points of the alkali metals also decrease down the group for the same reason.

All alkali metals show very distinctive colours when heated in a flame test. The colours range across the rainbow from the crimson flame of lithium to the blue-violet of caesium compounds.

Soft and not very dense

It is not only increased reactivity that arises from the outer electrons of the alkali metals roaming relatively free. In pure metallic form the electrons of these group 1 elements form a sea roaming freely among islands of positively charged ions. These mobile electrons are very good at conducting electricity and heat. It also follows that there are only loose attractions between each metal ion, shielded by the electrons, which makes these metals soft and easy to cut as atoms slip past each other. The far-flung outer electrons also increase the size of these atoms meaning that, in general, group 1 metals have the largest atomic radius of any other element in the same period. The smallest atoms are those on the right of the table as they do not want to lose any electrons and so hold on to them extremely tightly.

solutions when dissolved in water. An alkali is the opposite of an acid. Acids form solutions in water with an excess of hydrogen H+ ions swimming around, which are just hydrogen atoms with their electron removed. Alkalis, however, have a shortage of these H+ ions. Acids can be thought of as wanting to give away these H+ ions while alkalis want them. Many alkalis contain a hydroxide ion OH-, which is an oxygen atom bound to a hydrogen with an additional electron attached. If an acid is mixed with such an alkali then the H+ ion bonds with the OH- to form water: H_2O. The alkali metal ion left over then forms a salt with whatever negatively charged ion was originally

Lithium
One to watch

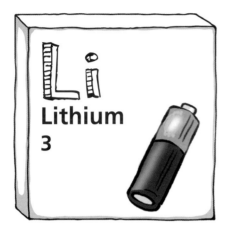

Atomic number	3
Atomic weight	6.942
Abundance	20 mg/kg
Radius	145pm
Melting point	181°C
Boiling point	1287°C
Configuration	(He) 2s1
Discovered	1817 – A. Arfwedson

Lithium is in a very exclusive club. Along with hydrogen and helium it is the only other element to have been created in the Big Bang. Our desire for electric cars and devices is also set to make it worth its weight in gold.

Red fire stone

In 1817 Johan August Arfvedson in Stockholm observed a deep crimson flame when he sprinkled the grey mineral petalite onto a fire. He immediately speculated that this mineral contained an unknown metal, naming it 'lithium' after the Greek for stone, *lithos*. Arfvedson tried to use electricity to extract pure lithium from a dissolved solution of the mineral in a process called electrolysis, invented by British chemist Humphry Davy (see Potassium). He was not successful. Lithium was instead first extracted from a molten salt of lithium chloride in 1855.

Lithium is a silvery-white metal that readily reacts with water but does not react with oxygen around us. Instead, lithium is one of only a few metals that readily reacts with the more common, but usually less reactive, nitrogen in air. If left for some time, the lithium nitride salt that is produced will start to stink. This is because lithium nitride reacts with the water vapour in the air to produce lithium hydroxide and smelly ammonia.

Highly reactive lithium is found predominantly in salt compounds on Earth but also in some other minerals. Sea water is estimated to contain over 200 billion tonnes of lithium as salt which makes up around 0.2 parts per million. Salt can

Lithium salts, like lithium carbonate, are used as mood-stabilising drugs.

be extracted from the brine solution through evaporation fields which leave it behind as a residue. The most accessible form of lithium, though, is from vast dried-up salt lake beds with the largest reserves found in South America, Chile in particular.

Relax and rewind

Lithium is moderately toxic to humans, as was discovered in the 1940s when deaths occurred from use of lithium chloride as a substitute for table salt. In small doses, though, it has been put to positive use. The calming effect of lithium carbonate was first shown in guinea pigs, when in 1949 the Australian doctor John Cade saw that they became relaxed and docile when injected with a dilute 0.5% solution of the salt. Straight away he saw the applications for his work in the Department of Mental Hygiene, and was soon injecting patients with the same solution In just days even some of the most severely troubled patients were able once more to resume their normal daily lives. Lithium is still used today around the globe to ease the suffering of those afflicted with bipolar disorder. It is not known exactly how lithium induces this calming effect, although it seems to prevent the excess production of certain chemical messengers in the brain.

Seeking riches

A new rush is on the horizon for the 21st century and it is not gold that prospectors want to dig up, it is lithium. With our unquenchable thirst for portable electronic devices and the birth of the electric car, all eyes have been on the prize of bigger and better batteries. The champion technology right now is lithium-ion batteries. Flow of electricity is just the movement of electrons through wires. As the most electropositive element, lithium is happy to lose its outer electron and become a positive Li+ ion, so it is a great source of electrons.

It is also the lightest element to form a solid at room temperature which means that a battery made from lithium would be ultra light. This double combination makes it fantastic for use in both portable devices and electric cars. With our consumption of electronic gadgets and the global aim for all cars to be electric in 50 years, the demand for lithium is set to grow and grow.

A new rush is on and it is lithium to make batteries, not gold, that prospectors are after.

Sodium

Keeps our body working

Na
Sodium
11

Atomic number	11
Atomic weight	22.9898
Abundance	23600 mg/kg
Radius	180pm
Melting point	98°C
Boiling point	883°C
Configuration	(Ne) 3s1
Discovered	1807 – H. Davy

The yellow light arising from the excitation of sodium's electrons is put to very good use in street lamps and fireworks. It is the most abundant of the alkali metals on our planet, making up around 2.6% of the Earth's crust by mass.

As it is a reactive alkali metal, no sodium is found in nature in elemental form; it is instead found combined with metals in salt compounds. Sodium chloride, for example, is common table salt, which we use to flavour food.

Used for centuries

Sodium compounds have been known of for centuries: natron washing soda was mentioned in ancient Egyptian hieroglyphs. Known today as sodium carbonate, it was used by the ancient Egyptians to make soap, and also in the process of mummification because of its water-absorbing and bacteria-killing properties. Natron also contributes its first two letter to give us the seemingly illogical symbol Na for sodium.

During the Middle Ages, Europeans used sodium carbonate to treat headaches. Western universities at the time were teaching the medicine of Islamic scholars and so the treatment adopted the name sodanum, derived from the Arabic for headache, *suda*. It was this that Humphry Davy was thinking of when he first extracted sodium in 1807 and named it (see Potassium for more).

Sodium does have a use in its pure elemental form because it is very efficient at transferring heat and finds use in liquid form to keep some nuclear reactors cool. With a melting point of 371K and a boiling point of 1156K it offers a massive 785K range of temperature over which it remains liquid. This is much greater than the range of 100K between water, ice and steam. To get larger liquid ranges in temperature water can be pressurised, but this adds extra safety concerns. Because sodium is a relatively heavy element is does not like absorbing neutrons, which is good because they are the particle that sustains the nuclear decay of Uranium in reactors. Absorbing too many neutrons

My great-grandfather just before leaving for the jungles of Burma (now Myanmar).

sodium it contains, is essential to life.

We have to take in around two grams of sodium each day to remain healthy, and more if we lose a great deal through our salty sweat. A fine balance between low sodium levels and high potassium levels inside cells is required for almost all processes in our body to happen. It is the exchange of sodium into and out of cells that regulates our body's messaging systems; from the slow release of hormones to the fast firing of nerve cells. It is also essential for the contraction of muscles with which we move, breathe and pump blood around our body.

Deadly sushi
A very effective way to kill someone, in fact, is to mess with this fine balance in the exchange of sodium into and out of cells. A chemical called tetradotoxin, TTX for short, will block the specialised channels that transport sodium into and out of cells. TTX can be found in certain parts of fugu (pufferfish), which if not carefully prepared could lead to death within minutes or hours of eating. There is currently no known antidote for TTX.

could extinguish the nuclear reactions and the core would not produce energy. Water, on the other hand, is quite good at absorbing neutrons so must be used sparingly to avoid nuclear reactions fizzling out. Of course sodium must be well contained because if it leaked then it would react explosively with the surrounding air. This considered, it still presents an overall lower risk than a number of other nuclear reactor designs.

Jungle trek
After sweating for days whilst trekking through the jungle in Myanmar during World War II, my great-grandfather suddenly passed out. When he regained consciousness he found a salty taste in his mouth and a fellow soldier offering him a block of salt to eat. Although often warned about links to high blood pressure and heart disease, my grandfather realised that day how salt, and the

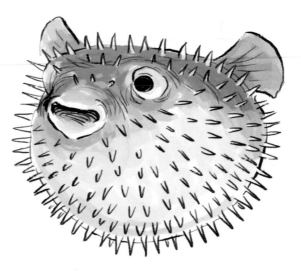

Fugu (pufferfish) is a delicacy that can only be prepared by master sushi chefs, otherwise it can be deadly.

Potassium
Pertinent to plants and people

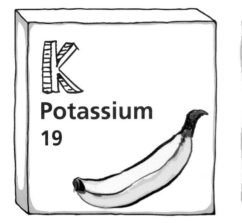

Potassium
19

Atomic number	19
Atomic weight	39.0983
Abundance	20900 mg/kg
Radius	220pm
Melting point	63°C
Boiling point	759°C
Configuration	(Ar) 4s1
Discovered	1807 – H. Davy

Potassium is so reactive it can burn a hole through solid ice or extract the oxygen from the fibres of a piece of paper. Yet this volatile element, like sodium, is essential for life on Earth.

Burn plants, dissolve the remaining ash in water and you'll find yourself with a solution of almost entirely potassium salts. This solution is called potash because the plant ash was mixed with water in a pot to produce it. Potassium was named after this solution by English chemist Humphry Davy after he isolated the metal from it in 1807. It was the first alkali metal and element to be discovered using the technique of electrolysis.

Electrolysis
Davy pioneered the procedure, which literally means unbinding (from the Greek *lysis*) by electricity (*electro*). The procedure splits apart ionic compounds using a positive electrical anode and a negative electrical cathode. A positive metal electrode, connected to an electricity supply, would attract the negative part of the compound. A negative metal electrode, connected to the same electricity supply, attracts the positive part of the

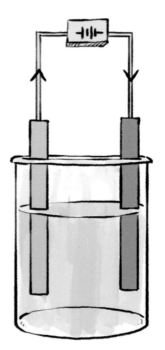

An electric current, passed through a molten or dissolved salt, attracts the positive metal ions towards the negative electrode. Here the metal ions gain electron(s) to become neutral atoms and form the elemental metal.

compound. Alkali metals prefer to lose an electron becoming positively charged, so are attracted to the negative-charged electrode. When they reach it, the positive ions pick up an electron to become a neutral atom.

When Davy performed this procedure with potash solution nothing happened. He persisted by making the potash more caustic, 'pure', by treating a solution of it with lime water which left him with a solution of potassium hydroxide. It was from this that Davy isolated the metal potassium using electrolysis. Edmund Davy, Humphry's young cousin and assistant for the experiment writes how when Humphry first saw 'the minute globules of potassium burst through the crust of potash, and take fire as they entered the atmosphere, he could not contain his joy'. Just months later, Davy isolated sodium using the same techniques.

Salty plants

While Davy preferred to link potassium to potash, the Swedish chemist Berzelius preferred the name kalium, because potash was prepared in the ancient world for bleaching textiles and making soap, predominantly from the herb kali. The name potassium remained, but because Berzelius invented the international system of chemical symbols we use today, it has the symbol K. The word alkali derives from the addition of the Arabic definite article 'al' to kali. This not only provides the name for the group 1 alkali metals but also all other chemicals with similar alkaline properties.

Radioactive banana

As is evident from potash remains, potassium plays an essential role in plant life. The most common use of potassium is in salts, usually combined with nitrogen and phosphorus, to make plant fertilisers. Bananas and tomato paste are among food products with the highest potassium content per gram and as such are also radioactive. Potassium plays an important role in animal life too, regulating many processes within cells (see Sodium). We require large amounts of potassium to regulate these processes which makes the rare potassium-40 the most common radioactive isotope to be found in our body.

Saltwort, known as kali in the ancient world, was burnt to produce potassium-containing potash.

Rubidium

Rb
Rubidium
37

Atomic number	37
Atomic weight	85.4678
Abundance	90 mg/kg
Radius	235pm
Melting point	39°C
Boiling point	688°C
Configuration	(Kr) 5s1
Discovered	1861 – R. Bunsen & G.R. Kirchhoff

Caesium was the first element to be discovered using flame spectroscopy and rubidium the second. The discoveries were made by German chemist Robert Bunsen and physicist Gustav Kirchhoff in 1861, when heating mineral water from a spring from Bad Dürkheim in Gemany.

Colourful connotation

Both elements derive their name from Latin descriptions of the flame colour they produce. Rubidium is derived from the Latin *rubidus*, meaning deepest red, and caesium after the Latin *caesius*, for sky blue.

The key to quantum

Rubidium was a key element in developing the quantum theory of particles. Unlike the electrons in orbit around an atom, when close to absolute zero in temperature, atoms can all occupy the same energy level, becoming indistinguishable from one another. This new state of matter (not solid, liquid, or gas) is known as a Bose-Einstein condensate, after the scientists who predicted its existence. The first of its kind was produced using super-cooled rubidium atoms (see Helium for more).

Antimatter tumour detector

Rubidium is not toxic to humans; once inside the body it is treated as if it were potassium and is rapidly lost in sweat and urine. This behaviour by the body has led to using the radioactive isotope rubidium-82 to locate brain tumours. Potassium, and therefore radioactive ^{82}Rb ions, collects in rapidly growing cancer cells. When the isotope decays, a flash of gamma radiation is released which escapes the body to be detected. This allows doctors to locate the exact position of a tumour.

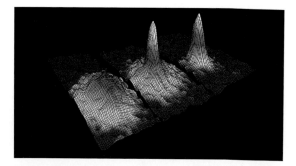

These plots show the energy distribution of rubidium atoms. From left to right the rubidium atoms begin to behave as one super atom as they cool down and enter a Bose-Einstein condensate state.

Caesium

Caesium
55

Atomic number	55
Atomic weight	132.90545
Abundance	3 mg/kg
Radius	260pm
Melting point	28°C
Boiling point	671°C
Configuration	(Xe) 6s1
Discovered	1860 – R. Bunsen & G.R. Kirchhoff

This element is all about the timing. It was discovered just before its lighter rubidium cousin, which it trumps as the connoisseurs' choice to make atomic clocks.

Keeping time

Both caesium and rubidium find use in atomic clocks, which use the transition of electrons between orbits of differing energy to accurately determine time. This is done by passing microwaves through metal vapour, which when tuned to the right frequency, causes all electrons to transfer to a higher energy level. This frequency remains the same the universe over, depending only upon the configuration of electron orbitals in the chosen atoms. The international definition of the second is defined as 9192631770 periods of this microwave radiation, corresponding to the transition between the two hyperfine levels of the ground state of the caesium 133 atom.

There is some jitter in the electrons' energy level as they are influenced by surrounding magnetic fields, but rubidium and caesium are the elements affected the least by this jitter. Caesium energy levels are less jittery and more stable than those of rubidium, which makes a caesium clock more accurate. The drawback of caesium clocks is their price: the rarity of caesium means that they cost around 700 times more than clocks made from relatively abundant rubidium.

Colourful relativity

Caesium is special because it is one of only three metallic elements which are not silver in colour; the others are gold and copper. Its colour is due to the effects of Einstein's special theory of relativity changing the energy of electron orbits.

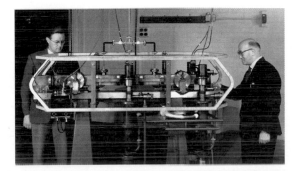

Physicists Jack Parry (left) and Louis Essen (right) adjusting the world's first caesium atomic clock, which they built in 1955 at the National Physical Laboratory in Teddington, UK.

Francium

Fr
Francium
87

Atomic number	87
Atomic weight	223
Abundance	$1\times10{-}18$ mg/kg
Radius	no data
Melting point	27°C
Boiling point	677°C
Configuration	(Rn) 7s1
Discovered	1939 – M. Perey

Element 87 was predicted by Mendeleev to fill a gap below caesium. With continued failure to find it, the search then turned to radioactive decays. Two rules dictate the change of elements through radioactive decay.

Alpha and beta

If a radioactive atom releases alpha particles then the new element will have an atomic number two lower (-2) than the parent element as it has lost two protons from its nucleus. If the element decays through beta decay on the other hand, then its atomic number increases by one (+1), as a proton is formed from a neutron while emitting an electron and electrically neutral neutrino particle. To fill the void of element 87 either an alpha-emitting isotope of actinium (89) was needed or beta-emitting radon (86).

The new generation

Radon only decays via alpha radiation, not via beta as would be required to produce element 87. Almost all, 99%, of actinium was seen to decay via beta decay creating thorium (90), an increased, not decreased, atomic number. Just 1% of the time, though, the actinium did decay via alpha decay which led to element 87. The difficulty was how to isolate this tiny amount, a job made all the more difficult by the short half life of the new element

87 of around 21 minutes. Into the limelight steps Marie Curie's protégé Marguerite Catherine Perey, who was able to extract the tiny fraction of this unstable element. Discovery of the element came in 1939, five years after Curie had passed away. Perey was encouraged by many in the field to take a degree and then a PhD. In 1946 she successfully defended her PhD thesis on the topic of the missing element 87, during which she also named the element 'francium' after her home country. Sixteen years later, she was the first woman to be elected to the French Academy of Sciences, an honour not even bestowed upon Marie Curie herself.

French nuclear chemist Marguerite Perey (1909–75), discoverer of Francium and a former student of Marie Curie.

Alkaline Earth Metals

The alkaline earth metals of group 2 are a collection of reactive metals that are not found in elemental form in nature.

Each element has two valence electrons filling their outer s subshell, which makes them a little more stable and slightly less reactive than group 1. Every element in this group is found in nature although very little radium exists as it is created only as part of a decay chain of heavier elements.

Ionic extraction

Group 2 elements react with non-metal halogens of group 17 and create salt compounds that form alkaline solutions if dissolved in water. All but beryllium form ionic salts, and it was in the electrolysis of these molten salts that the elements were first isolated. Beryllium, however, forms covalent bonds and is extracted via a complicated sequence of chemical extractions.

Windows to unseen light

All of group 2 form fluoride compounds that are insoluble, i.e.they cannot be dissolved in water. These fluorides are also transparent to more than just the visible light we see, but also higher-energy ultraviolet and infrared. Windows made from group 2 fluorides are used in infrared spectroscopy as they absorb very little of the light emitted by other compounds. Calcium fluoride is the most common material for such windows, but the more expensive barium fluoride is used if viewing lower energies of light; the heavier barium atoms vibrate less than lighter calcium and so it requires longer wavelengths of light to set them in motion.

Group 2 elements demonstrate the same trends as seen in group 1 with reactivity and softness increasing with atomic radius down the group.

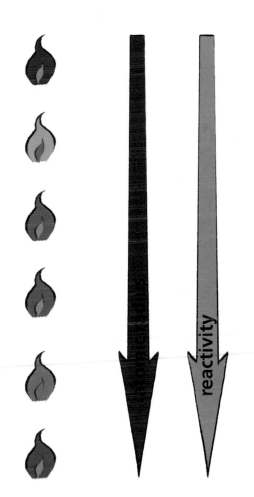

The vivid range of colours that the alkaline earth metals show when excited by a flame. As the atomic radius increases down the group, so too does the reactivity and softness of these metals.

Beryllium
Lending lightness and strength

Be
Beryllium
4

Atomic number	4
Atomic weight	9.0122
Abundance	2.8 mg/kg
Radius	105pm
Melting point	1287°C
Boiling point	2469°C
Configuration	(He) 2s2
Discovered	1797 – L.N. Vauquelin

Most elements lighter than iron are cooked up in the nuclear fusion factory at the centre of stars (see Hydrogen) but stable beryllium is not. This makes it the second rarest element lighter than iron.

The stable beryllium-9 isotope is only produced in interstellar space when heavier elements are split apart by cosmic rays. Cosmic rays are high-energy charged particles which fly through space; they are bombarding the Earth's atmosphere right now from all directions.

A magnetic history

Beryllium-10, a radioactive isotope of beryllium, is formed on Earth through collision of cosmic rays and stable beryllium-9. The magnetic field of our planet protects us against these cosmic rays but its protection wavers. At times in history the magnetic field of the Earth was strong and protecting but at

The Earth's magnetic field continually protects us from a barrage of charged particles known as cosmic rays.

other times it has been weak. A weaker magnetic field would allow more cosmic rays to flood the Earth's surface, creating more beryllium-10. Through measuring the amount of radioactivity from beryllium-10 in deep ice core samples, scientists can map out the historical strength of the Earth's magnetic field over millions of years.

Fuselage, fuel and fatality

Beryllium's rarity on Earth affords it no known biological role. It is lightweight, mechanically strong and releases huge amounts of heat when burnt. For these reasons it was hailed in the 1950s by the aerospace industry as a wonder construction material and fuel for aeroplanes. Despite this seemingly promising future, only 500 tonnes of the metal are today extracted each year, because beryllium is extremely toxic. Exposure to beryllium dust wreaks havoc in humans, causing chronic inflammation of the lungs and shortage of breath. This condition, known as berylliosis, can take up to five years to manifest, with around a third of those affected dying prematurely and the rest left permanently disabled.

Beryllium windows such as the dull grey one in the very centre of this disc are used in high- or low-pressure laboratory equipment as they are transparent only to high-energy particles or X-rays.

Despite this, beryllium is still used in manufacturing. With just five neutrons, four protons and four electrons, beryllium is so light that it is ignored entirely by high-energy particles and radiation. A beryllium window is more transparent to X-rays and gamma rays than one made of glass. These windows are used to cover the 'open' ends of specialist equipment where you wish X-rays to come out. They are also used at the Higgs boson-discovering Large Hadron Collider at CERN, where protons are accelerated to near the speed of light. To achieve such a feat protons must travel unhindered through tubes evacuated of all air. The ends of these vacuum tubes are sealed with beryllium windows, which stop air from getting in while also allowing the high-energy protons to escape unscathed.

The neutron

A beryllium window also played an essential part in the discovery of the neutron. In 1932 James Chadwick bombarded a sample of beryllium with alpha particles emitted from radium. He observed a new kind of subatomic particle being released, which had mass but no charge. The radium-beryllium combination is used to this day to generate neutrons for research; it requires around a million alpha-particles to produce about 30 neutrons.

Attractive

Beryllium does not ionically exchange its two valence electrons but instead shares them with halogen atoms, forming a covalent bond. If a beryllium 2+ ion were formed it would have an incredibly high density of electric charge. The force exerted by this ion would be great enough to distort the electron orbital clouds of other atoms to the point where they overlapped with its own. Overlapping orbitals are how electrons are shared and covalent bonds formed. Compounds of beryllium are consequently not very good conductors of electricity, which added to the difficulty in isolating the metal.

Magnesium
Harnessing sunlight

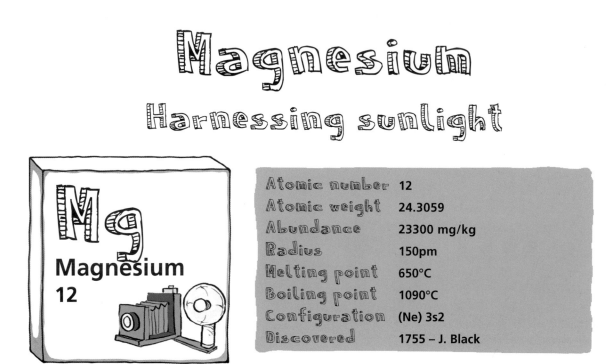

Atomic number	12
Atomic weight	24.3059
Abundance	23300 mg/kg
Radius	150pm
Melting point	650°C
Boiling point	1090°C
Configuration	(Ne) 3s2
Discovered	1755 – J. Black

In the summer of 1618, it was noticed by a man named Henry Wicker that despite a drought, his cows refused to drink from a pool of water on Epsom Common, UK. Wicker tasted the water, found it bitter and took a sample home. On evaporating the water he was left with a salt which, as he had already experienced, acted as a laxative. These 'Epsom Salts', which comprised the compound magnesium sulphate, were used for the next 350 years to treat constipation.

Energy from the sun to our cells

With this kind of effect on our lower bowels you might be surprised to learn that magnesium is an essential ingredient for life on Earth. Magnesium sits at the centre of the green-looking complex chemical chlorophyll, providing it with its essential unique shape. This complex molecule is used by plants and other life forms to convert carbon dioxide gas and water into glucose, using the energy of sunlight. Glucose is then combined

Magnesium is at the heart of the chlorophyll molecule which harnesses sunlight to be used in photosynthesis of glucose sugar.

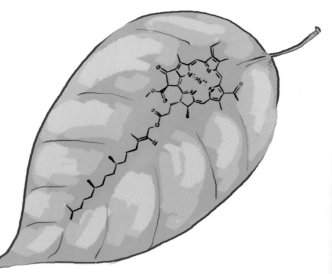

Light and strong magnesium is used to give structure to everyday hand-held electronics.

to form starch, cellulose and lots of other tasty molecules for consumers like us to eat. These chemicals are then broken back down along with oxygen in our bodies to form carbon dioxide and water once more, in a process called respiration. The energy released in this process is used to form another chemical in our bodies called ATP. ATP is used to transfer energy around the body and release it where and when needed (see Phosphate). To create ATP in this process requires the presence of another magnesium-containing chemical.

Lightweight construction

Magnesium is also the lightest metal which can actually be used; not as reactive as lithium or sodium, and far from as toxic as beryllium. It is also one of the most abundant elements in the Earth's crust, sixth by weight, making it readily available. It is relatively strong and is used to structure lightweight electronic devices such as mobile phones and laptops. Magnesium has also been used for the hulls of boats, the skins of aeroplanes and high-performance car parts.

Reacting

Magnesium is still a moderately reactive metal, burning with an intense white light in air. This white light has historically been used in flash bulbs for photography. It is also one of only a handful of metals that will react with nitrogen. The metal is prevented from reacting with oxygen, water or nitrogen because it quickly forms a protective outer magnesium oxide coating.

Magnesium's central role in biological processes led many scientists to attempt to create organic magnesium compounds. The organomagensium chemicals were produced in the late 19th century but none could be dissolved in water, meaning they were not suited to propagate organic reactions. In 1900 a young French PhD student Victor Grignard had an new idea; he reacted various organic halides with magnesium in ether. This led to stable solutions of organomagnesium which could be used for organic reactions, now named Grigand reagents. After the first publication of this process organic chemists around the world immediately applied his procedure. He was awarded the Nobel Prize in Chemistry in 1912, and today over 100,000 papers have been written on organomagensium Grignard reagents.

Magnesium is added to sparkler fireworks to provide the bright white sparkle as it burns.

Calcium

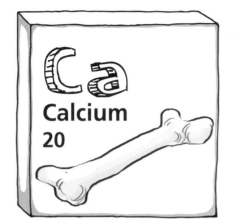

Atomic number	20
Atomic weight	40.078
Abundance	41500 mg/kg
Radius	180pm
Melting point	842°C
Boiling point	1484°C
Configuration	(Ar) 4s2
Discovered	1808 – H. Davy

Calcium is all around us: in the water we drink, the concrete our homes are constructed with and the rock upon which they are built. Calcium is the fifth most abundant element in the Earth's crust and one which has become essential to life.

As adverts regularly remind us, the best source of calcium in our diet is from dairy products. Milk has evolved to give offspring all of the essential minerals they require. Calcium is used by the body to produce a number of minerals in a process called biomineralisation.

Bones and teeth

An average adult human will have over 1kg of calcium in their body, 99% locked up within bones. Calcium phosphate is combined with organic molecules to create bones and teeth. By varying the mixture, our body produces materials with different strengths and flexibility. Our bones function not only as a structural support but also a central calcium repository. During pregnancy a mother's bones are raided for the mineral in a process called demineralisation. As we grow older our bones also lose much of this calcium which can lead to brittle bones: osteoporosis. Older people are therefore more susceptible to breaking a bone under stress.

Super shell-seekers

Calcium carbonate is used by molluscs and other sea creatures to make protective shells. Hermit crabs do not make their own shells but instead inhabit second-hand shells, changing to larger shells as they grow. Transition between shells is a dangerous time, leaving them vulnerable to being eaten by predators. Some hermit crabs have evolved to minimise this risk with an ability to recognise shells by measuring minute amounts of calcium carbonate dissolved in the water. Hermit crabs have been reported to have an amazing sensitivity to calcium of the order less than 4 atoms of calcium among a million others!

Hermit crabs are able to sense the tiniest amount of calcium to help them find a new home as quickly as possible.

The secret life of the periodic table

Strontium

Sr
Strontium
38

Atomic number	38
Atomic weight	87.62
Abundance	370 mg/kg
Radius	200pm
Melting point	777°C
Boiling point	1382°C
Configuration	(Kr) 5s2
Discovered	1755 – A. Crawford

Strontium does not seem to form the same variety of minerals as the other group 2 elements. Despite being the 15th most abundant element in the Earth's crust it is mainly found either as celestite (strontium sulphate) or strontianite (strontium carbonate).

Strontianite was named after the village of Strontian in the Scottish Highlands and gives the element its name, making it the only element named after a location in the UK.

Flat screen flop

At the turn of the 21st century strontium oxide was added to the glass screens of old-style, cathode ray TVs as it blocked X rays emitted from the TV's tube. Since the advent of flat screen TVs, strontium is used mainly to provide the deep red colour of fireworks and flares.

Vegetarian warriors

Similar in size to their calcium counterparts, strontium ions are regularly absorbed by the body, most finding their way into our bones and teeth. Due to the continuous uptake of minerals, plants contain a larger amount of strontium than animals. Forensic archaeologists have determined that Roman gladiators were vegetarian because of large strontium deposits in their bones.

Lots of strontium in the bones of Roman gladiators suggests they had a primarily vegetarian diet.

A dangerous past

The radioactive isotope strontium-90 is a product of the decay chain of uranium. Cold War detonations of atomic bombs released large amounts of this isotope into the atmosphere, and it was seen to have biomineralised in the milk teeth of American children. As with accumulation of any radioactive material, there is an increased risk of cancer; as seen following the Chernobyl nuclear reactor disaster in 1986. Strontium's bone-seeking ability has led to a less sinister medical use, using radioactive strontium-90 in small doses as a radio therapy to target bone cancer.

Barium

Atomic number	56
Atomic weight	137.327
Abundance	425 mg/kg
Radius	215pm
Melting point	727°C
Boiling point	1897°C
Configuration	(Xe) 6s2
Discovered	1808 – H. Davy

This heavyweight puts its size to good use in drilling for oil and checking our bowels.

Despite being more abundant than strontium; barium fetches a much higher price because it is desired for its weight. Minerals containing barium are much heavier than other alkaline earth metals as it is the highest density stable element in the group; which is also the root of its name, from the Greek word *barys*, which means heavy. Dominant use of barium is in the form of its most common mineral barite (barium sulphate). This heavy mineral increases the density of fluids used by gas and oil companies when drilling to prospect for new supplies.

Barium sulphate is also used to give us a glimpse inside our body. As a metal and as an ion barium is toxic to life, causing heart trouble, tremors and paralysis. Despite this, a large amount of barium sulphate is swallowed down the mouths, or injected up the other end, of hospital patients each year. Barium sulphate is non-toxic because it cannot dissolve in water where it would free the dangerous barium ions to wreak havoc in our body. Heavy barium, unlike light beryllium, is very good at scattering X-rays. The soft tissues of our digestive system are made from light elements and also pretty much invisible to X-rays. Like bone, however, heavy barium scatters X-rays, illuminating the otherwise invisible digestive system as it travels through. These foul-tasting 'barium meals' are usually mixed with strawberry or mint flavouring to make them more palatable, sadly with little effect.

Heavy barium sinks into drill holes aimed at tapping reserves of fossil fuels.

The secret life of the periodic table

Radium

Ra

Radium
88

Atomic number	88
Atomic weight	226
Abundance	9x10–9 mg/kg
Radius	215pm
Melting point	700°C
Boiling point	1737°C
Configuration	(Rn) 7s2
Discovered	1898 – P & M. Curie

An element that played a shining role in modernising labour law.

Named after its high radioactivity, radium was found with another radioactive giant, polonium, in 1898. Uranium salts extracted from the mineral pitchblende were used in 19th-century Bavaria to colour pottery glaze. Pierre and Marie Curie noticed that the waste left from this process remained highly radioactive; sifting through tonnes of it revealed these two new chemical elements.

Treating cancer
In Marie Curie's lifetime radiotherapy was used in hospitals to treat cancers; radioactive radon gas released from the decay of radium killed rapidly dividing cancer cells. Its use also sparked the fad that water treated with radium promoted good health. Radioactivity is not discriminant though, also killing healthy cells in the process. This dark nature was all too present in the story of the 'radium girls'.

Rewriting law
During World War I a company called US Radium made and supplied watches to the US military, with dials illuminated by radioactive radium paint. The ladies who painted these dials often licked their brushes to ensure an accurate and thin stroke. Workers soon came out in sores, suffered cancers of the mouth or died from a radiation-related illness. US Radium urged medical professionals to attribute death to other causes; syphilis was quoted in some cases to discredit the women in question. In 1928 a group of the workers eventually found a lawyer willing to represent their case to the courts. The case was successful and rewrote labour laws and instigated safety standard regulations in the US.

Factory workers painting fine, radium paint, glowing lines onto dials and clocks.

Transition Metals
Colourful catalysts of different designs

The transition metals each have a partially filled d subshell which can accommodate up to ten electrons in total. The chemistry of these elements is all about give and take of electrons.

If an element loses an electron in a chemical reaction, to form ionic bonds, it is said to have been oxidised. The count in the number of electrons exchanged is known as an ion's oxidation state. Positive oxidation states relate to an atom having lost electrons to become a positive ion. When a negative oxidation state is assigned it is because the ion has gained an electron. An oxidation state of zero is given to neutral atoms, where the number of electrons perfectly balance the protons in the nucleus.

Catalysts

The d block electrons are loosely held by most transition metals. They are happy to lose, and in some cases accept, electrons and therefore have

The various oxidation states of the transition metals provide them with colourful and catalytic uses.

Z	name	-5	-4	-3	-2	-1		+1	+2	+3	+4	+5	+6	+7		group
19	potassium					-1	K	+1								1
20	calcium					-1	Ca	+1	+2							2
21	scandium						Sc	+1	+2	+3						3
22	titanium				-2	-1	Ti	+1	+2	+3	+4					4
23	vanadium			-3		-1	V	+1	+2	+3	+4	+5				5
24	chromium		-4		-2	-1	Cr	+1	+2	+3	+4	+5	+6			6
25	manganese			-3	-2	-1	Mn	+1	+2	+3	+4	+5	+6	+7		7
26	iron		-4		-2	-1	Fe	+1	+2	+3	+4	+5	+6			8
27	cobalt			-3		-1	Co	+1	+2	+3	+4	+5				9
28	nickel				-2	-1	Ni	+1	+2	+3	+4					10
29	copper				-2		Cu	+1	+2	+3	+4					11
30	zinc				-2		Zn	+1	+2							12
31	gallium	-5	-4		-2	-1	Ga	+1	+2	+3						13
32	germanium		-4	-3	-2	-1	Ge	+1	+2	+3	+4					14
33	arsenic			-3	-2	-1	As	+1	+2	+3	+4	+5				15
34	selenium				-2	-1	Se	+1	+2	+3	+4	+5	+6			16
35	bromine					-1	Br	+1		+3	+4	+5			+7	17
36	krypton						Kr		+2							18

the ability to take on a wide variety of oxidation states. In the very middle of the d block, in period 4, is manganese. With its half-filled shell it is able to take on an impressive ten different oxidation states from -3 right through to +7. The ability of an element to take on this vast number of guises aids all manner of chemical reactions. Transition elements exchange electrons between other atoms to ensure reactions occur efficiently and quickly. They do this without becoming products of a reaction themselves: substances which show these properties are called catalysts. Transition metals catalyse everything from the making of plastics to the safe disposal of toxic gases.

Conductors

The free movement of electrons also leads to some of these elements being very good conductors of heat and electricity. Electricity is simply the movement of electrons en masse in a particular direction; the freer they are the easier they are to move. Transfer of heat is more effective if energy is spread through light, fast-moving electrons than slower, bulky ions or atoms. Another property is magnetism, which arises essentially from all electrons being able to face in the same direction – free electrons can choose where they look. Electrons are also easily transitioned between different energy levels in these atoms. This produces a fantastic variety of colours which change dependent upon oxidation state. Rainbow spectra of compounds are created, but these elements also add colour to fireworks, flares and electric lighting.

Confusion

There is not a single way to perfectly label this collection of metals. In this book we have chosen to follow the popular definition that all d block metals, in groups 3 to 12, are transition elements. The International Union of Pure and Applied Chemistry holds a different definition, stating that transition metals are all elements which have, or can form negative electrically charged ions with a partially filled d shell. With a full d shell group

12 elements are therefore not strictly transition elements, which shows in their very different characteristics. Joining the transition elements following this definition would be scandium and yttrium because they too have partially filled d shells in their metallic state. These two elements, however, do not show the characteristic catalytic properties of the other transition elements. The lanthanides and actinides are also transition elements as they too bridge the gap between p shell and d shell valence elements. Some periodic table representations extend the transition to include these 'inner transition metals'. This 32-column and group table also better represents the configuration of electron shells in the atom and the rules used to fill them.

There are multiple definitions of what a transition metal is, some adding metals to this collection and others taking them away.

Scandium

Sc

Scandium

21

Atomic number	21
Atomic weight	44.9559
Abundance	22 mg/kg
Radius	160pm
Melting point	1541°C
Boiling point	2836°C
Configuration	(Ar) 3d1 4s2
Discovered	1879 – F. Nilsen

Mendeleev's missing element: providing the next generation of fuel tank?

In his second periodic table of 1871, Mendeleev predicted the existence of a metal between calcium and titanium, with an atomic weight of around 44, that would combine two of its atoms with three of oxygen. Just ten years later, Swedish chemist Lars Nilson identified a new metal in an ore from unique spectral lines, naming his element scandium, from the Scandinavian region of which Sweden is a part. It was fellow Swedish chemist Per Theodor Cleve who recognised this element as Mendeleev's missing metal and informed him of Nilson's discovery. Scandium has molecular weight 45, and forms scandium oxide, Sc_2O_3, as Mendeleev predicted. Nilson had isolated this oxide; the elemental metal was only extracted in experimental-sized quantities over a century and a half later in 1937.

Everywhere
As it has no geological role, scandium is found widely distributed in many different ores across the globe. This means that although similar in abundance to lead, it cannot be mined, and is found only as a by-product of other metal ore mining. Scandium is limited to the +3 oxidation

The holes in this scandium sheet molecule are perfect for soaking up hydrogen, storing it to be used later as fuel.

state, making its chemistry less diverse than some of its transition metal cousins. The metal is very light and can be added to other metals to form light and strong alloys for making bike frames, although it cannot compete with the cheaper carbon fibre and titanium alloys on cost.

Energy future
Scandium is set to find a use though – to create lightweight storage tanks for hydrogen-fuelled cars. When combined with organic molecules, scandium metal can form porous materials which have large spaces like a sponge. These materials can soak up cold hydrogen and release it at a later time when they are heated. Such materials would safely and efficiently store hydrogen and do away with the need for heavy thick metal tanks that currently store the gas under massive pressure.

Titanium

A true titan of technology

Ti

Titanium

22

Atomic number	22
Atomic weight	47.867
Abundance	5650 mg/kg
Radius	140pm
Melting point	1668°C
Boiling point	3287°C
Configuration	(Ar) 3d2 4s2
Discovered	1791 – W. Gregor

Titanium is difficult to isolate, but we couldn't do without this element or its compounds in today's modern world.

Titanium is common on Earth but very difficult to extract from its strongly bonded compounds. Pure titanium metal is extracted today via the Kroll process from the most common titanium compound, titanium dioxide (TiO_2), which can be found in large deposits across the globe. The titanium dioxide is heated along with carbon to about 1000°C before chlorine gas is passed over it. A reaction produces titanium tetrachloride ($TiCl_4$), which is known in the industry as 'Tickle'. The Tickle is kept in an atmosphere of argon gas, as it would happily react with oxygen or water in the air to reform TiO2 once more. In this inert atmosphere, however, it reacts with hot and more reactive (and cheaper to acquire) magnesium at 850°C to produce the metallic element.

Titanic constructions

Once produced, the titanium metal forms a thin protective coating of TiO_2 over its entire surface, preventing it from reacting with the air any further. Extraction is expensive and so the lightness and strength of titanium metal is used in situations

The SR-71 Blackbird, the fastest jet ever built, was made predominantly from light and strong titanium.

where cost is not a factor. Titanium is used in the production of lightweight military and civilian aircraft, including the world's fastest manned jet, the SR-71 Blackbird. It is also used in high-end watches and glasses, and for jewellery. The oxide-coated metal doesn't react with seawater either, so finds use as a material for boat and submarine propeller shafts where strength and lightness are essential.

Titanium is non-toxic and also connects with bone so has found use in high-end joint replacements, especially the replacement of hip joints. Titanium is also an excellent catalyst in a number of reactions between organic compounds: only a tiny amount is required to create tonnes of plastics in a process called polymerisation. In a lot of polythene products that you buy, there is a trace amount of titanium which was used as a catalyst to connect smaller molecules into long strands.

Compound designs

Titanium dioxide, from which the metal is extracted, is relatively cheap and also has a huge range of uses. The white compound is used to lighten everyday products from toothpaste to paint; you are probably sitting in a room surrounded by TiO_2 right now. It is also used in the food industry to whiten sweets, cheeses and icing as additive number E_{171}. Titanium dioxide is also very good at absorbing ultraviolet (UV) light, which proves useful in sunscreen. The UV light liberates electrons from the TiO_2 when it is absorbed, which can go on to react with organic molecules with devastating effect. To make it safe for use in sunscreen it is usually coated in a protective layer of silica or alumina which absorb these free electrons. In other situations, however, these dangerous free electrons have their uses. Research is taking place into producing tiles with a thin surface coating of TiO_2, which could be used in hospitals to kill germs. Water cannot form droplets on TiO_2 and instead disperses, which means that dirt and other water-based residues are not left behind. TiO_2 may also find a use to coat construction materials, keeping exteriors of buildings, public foot paths and roads cleaner.

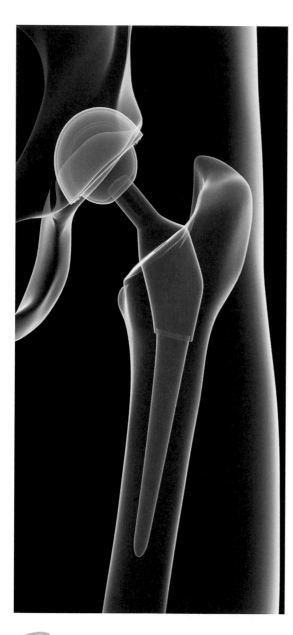

A hip replacement made from titanium metal. Thanks to a robust layer of oxide covering the metal, titanium does not corrode, which makes it perfect for use in the harsh chemical conditions inside the human body.

Vanadium

Vanadium
23

Atomic number	23
Atomic weight	50.9415
Abundance	120 mg/kg
Radius	135pm
Melting point	1910°C
Boiling point	3407°C
Configuration	(Ar) 3d3 4s2
Discovered	1801 – M. del Rio

Confusion and doubt surround both the discovery of vanadium and its role in life processes.

German chemist Baron Von Humboldt was reputedly the first to discover vanadium over 200 years ago, but both his samples and his notes were lost in a shipwreck. Following this, vanadium was discovered by several famous scientists, but it was not until 1831, when the Swede Nils Sefstrom was the first to gain sufficient and convincing evidence of its existence, that the discovery was taken seriously.

Colourful character
Sefström named vanadium after the Norse goddess Vanadís, associated with beauty and fertility, because of the many beautifully coloured chemical compounds it produces.

The rainbow colours of vanadium compounds and their chemistry arises, like that of all transition metals, from vanadium's variety of oxidation states: seven in total from -1 to +5. Its colours are ideal for spectroscopy, where they are used to identify the active parts of biological catalysts; enzymes.

Protect and prevent
Vanadium finds some biological use; exactly what is a hotbed of debate. Some marine animals such as sea squirts and the fairytale toadstool *amanita muscaria* (which is actually a mushroom) collect large amounts of vanadium, but it is not clear why. It may be to poison possible predators or to protect more sensitive molecules in the organisms from being split apart by hydrogen peroxide, which is readily broken down in the presence of vanadium. One clear biological property of vanadium ions is an ability to enhance, but not mimic, the action of the hormone insulin. After successful experiments, a number of vanadium compounds are being trialled in humans as potential treatments for diabetes: a disease in which people suffer high blood sugar levels due to a reduction in the amount or effectiveness of the hormone insulin.

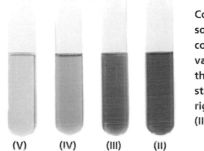

(V)   (IV)   (III)   (II)

Colourful solutions containing vanadium in the oxidation state (left to right) (V), (IV), (III) and (II).

Chromium

Cr
Chromium
24

Atomic number	24
Atomic weight	51.9961
Abundance	102 mg/kg
Radius	140pm
Melting point	1907°C
Boiling point	2671°C
Configuration	(Ar) 3d5 4s1
Discovered	1798 – L.N. Vauquelin

A rainbow of colours that add gemstone sparkle to otherwise colourless minerals.

Shiny chrome plating prevents the metal below from corroding.

When I was growing up, the word 'chrome' brought to mind the shiny metal bumpers and alloy wheels of cars. Chromium metal shines through a thin layer of chromium oxide, which prevents any metal underneath from corroding through reaction with air. Elemental chromium metal has very few uses beyond plating metal or the production of stainless steel.

Colourful salts

Chromium was named after the Greek for colour (*chroma*) because of its many colourful compounds; this was despite the objection of its discoverer, French chemist Louis Vauquelin, who pointed out that the metal itself has no inherent colour. The colourful compounds of chromium range from the dark red chromium oxide (CrO_3) to the violet anhydrous chromium (III) chloride ($CrCl_3$). They have been used for centuries as pigments and dyes. Chromium (IV) oxide (CrO_2) is magnetic and used to manufacture magnetic tapes for data backup and storage. Chromium oxide compounds are also used to make catalysts for making half of the world's polyethylene.

Sparkle and bling

The most exciting display of chromatic colour is in gemstones. With just a hint of chromium, the colourless minerals corundum, beryl and crysoberyl are transformed into the gemstones of ruby, emerald and alexandrite. Alexandrite is the most fascinating because its colour changes dependent upon the direction from which you look at it, i.e. it is pleochroic.

Manganese

Mn
Manganese
25

Atomic number	25
Atomic weight	54.938
Abundance	950 mg/kg
Radius	140pm
Melting point	1246°C
Boiling point	2061°C
Configuration	(Ar) 3d5 4s2
Discovered	1774 – J.G. Gahn

Despite being the fifth most abundant metal in the Earth's crust, many people confuse manganese with the more familiar magnesium. The confusion arises from their common origin in ores from the Magnesia region in northern Greece.

In the 17th century, the term 'white magnesia' was adopted for magnesium minerals and 'black magnesia' for the darker manganese oxides, which were used by stone age man to create cave paintings. We also owe this region the word 'magnet' which comes from the magnetic ore magnetite, also found here. Manganese metal is not magnetic itself, but bizarrely the colourless salt manganese sulphate ($MnSO_4$) is. The reason for this is that the 5 d shell electrons of the manganese in this compound are all unpaired, and so all of their spins can align with any magnetic field.

Deinococcus radiodurans **is the toughest extremophile bacteria known. It can survive up to 3,000 times the radiation dose that would normally kill a human thanks to the protection of a manganese-containing enzyme.**

Rolling steel

Around 90% of manganese metal that is produced from ore is used in the steel industry. When steel was rolled or forged, early industry pioneers found that it would break up. Englishman Robert Forester Mushet discovered that adding small amounts of manganese to the molten iron solves the problem. This is because manganese removes sulfur impurities from the steel due to a greater desire than iron to bond to sulfur. It converts low-melting iron sulphide in steel to high-melting manganese sulphide.

Cell protection

Our cells are constantly repairing our DNA against attacks from chemical free radicals such as the superoxide O_2^-. Free radicals are extremely reactive ions that split apart large organic molecules with ease. Manganese, as part of the enzyme manganese superoxide dismutase (Mn-SOD), converts O_2^- to safer hydrogen peroxide (H_2O_2). This process prevents our cells from being inundated with DNA repairs.

Iron

The end of a star

Atomic number	26
Atomic weight	55.845
Abundance	56300 mg/kg
Radius	140pm
Melting point	1538°C
Boiling point	2861°C
Configuration	(Ar) 3d6 4s2
Discovered	5000 BCE

Stable but reactive iron plays its part in the cosmic story of Earth and life on it.

A ball rolled down the side of a valley will release energy as movement, eventually coming to rest at the bottom where it is most stable and has least energy. Everything in nature wishes for the same; to be as stable and low in energy as possible. Chemistry is concerned with electric charges and the electromagnetic (EM) force that governs their behaviour. Inside every atom, however, there are other forces at play. Neutrons and protons in the nucleus are attracted to other protons and neutrons through the attractive strong nuclear force; without it no stable atomic nucleus could exist.

Keeping a nucleus together

Positive electrically charged protons constantly push away from each other because of the EM force. It is the pulling strong force which keeps them together against the will of the EM force. Stability of a nucleus is, therefore, a tug of war between the EM force trying to tear it apart and

the strong force binding it together. Neutrons do not change the chemistry of an atom as they have no charge, but they are essential to form stable atomic nuclei. They sit beside and space out protons which reduces the strength with which the EM force pushes them apart. Neutrons also provide the nucleus with additional attractive strong force.

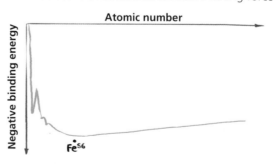

Of the two forces at play in a nucleus, it is the binding strong force that dominates the most in atoms of iron. In all other atoms, the electromagnetic force pushing the protons apart fights back to a greater extent, making those nuclei less stable.

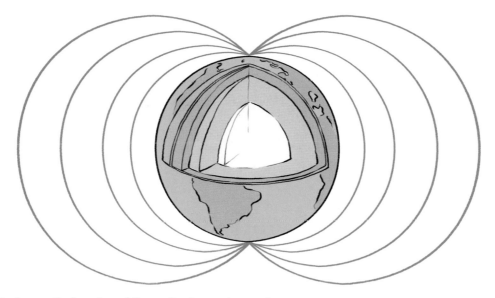

The circular (convective) motion of the molten iron outer core is thought to produce our planet's protective magnetic field.

Building up to it

Nuclear fusion within a star releases energy because the heavier nuclei it creates are more stable than their lighter components. But like the ball in the valley, there is a bottom lowest energy and greatest stability of nuclei. Iron may be chemically reactive but it has the most stable nucleus of all the elements. If you tried to fuse anything to an iron nucleus you would not gain any energy in return; it would require energy. This stability means that iron is the heaviest element that a star can cook up in nuclear fusion. As the end point of a star's life iron is abundantly found in the universe compared with most other metals.

Central to our planet and life

Iron is the most common element on Earth as it makes up almost the entirety of the inner and outer core of our planet. It is highly magnetic due to a number of loosely held and easily manipulated valence electrons. It is movement of iron in hot liquid form within the outer core that is responsible for the Earth's protective magnetic field.

The same valence electrons also allow iron to take many different oxidation states. This property makes iron amazing at moving electrons around, a property used by life on our planet in a number of different ways. Iron is at the heart of numerous enzymes which speed up processes within cells. Iron is also at the heart of haemoglobin and myoglobin. These proteins, found in our red blood cells, form complexes with molecules of oxygen and allow it to be transported around our bodies. The proteins then release the oxygen where it is used by cells to release energy in the process of respiration.

Getting old and forgetful

Some oxidation states of iron (iron III) are not soluble in water. If not part of an enzyme or transport protein, this form of iron precipitates out as a solid. On the occasion a cell is overrun by lots of solid iron it has to be destroyed, as there is no real way of recapturing it. Neurons, the nerve cells from which our brain is constructed, form unique connections which map out our thoughts and feelings. Types of degenerative disease, such as Alzheimer's, are brought on by this slow accumulation of iron in cells over many years, rendering neurons inactive.

Cobalt

Co
Cobalt
27

Atomic number	27
Atomic weight	58.9332
Abundance	25 mg/kg
Radius	135pm
Melting point	1495°C
Boiling point	2927°C
Configuration	(Ar) 3d7 4s2
Discovered	1739 – G. Brandt

Cobalt is difficult to extract from its ore and poisonous arsenic oxide often accompanies the process. These characteristics were both attributed by early miners to the work of evil spirits, and the element's name derives from the German *kobold*, meaning 'goblin' or 'evil sprite'. Cobalt has been used since ancient times as a pigment, for example in Egyptian blue paints and Greek glass vases.

Tricky and dangerous to extract, cobalt owes it name to *kobold*; German for goblin.

Hard to find

Cobalt is the lightest of the elements that cannot be produced in the heart of a star but must instead be produced in its explosive death. Large enough stars will end their lives in a supernova explosion where the majority of the star is violently expelled into space. The energy involved in this process is enough to fuse lighter elements into heavy elements that are impossible for a star to make. Cobalt and elements with a greater atomic number were formed in this process. For this reason, cobalt is 2500 times less abundant than neighbouring iron and is only found in large quantities along with other transition metals. It is usually sourced as a by-product of copper mining.

Magnetic motivations

Cobalt is chemically similar to iron and, along with nickel, is one of only three ferromagnetic transition metals elements. These elements can form permanent magnets and are also attracted by magnetic material. Cobalt finds use in the reading and writing of magnetic tapes and computer disks, speakers and electric motors. The metal remains magnetic to a higher temperature than other metals and cobalt alloy high-temperature magnets are used in turbo machinery motors. Cobalt also lends its high melting point to its alloys to create superalloys which retain their strength up to high temperatures. Such alloys are used to coat drills, saws and aircraft turbines.

Nickel

Atomic number	28
Atomic weight	58.6934
Abundance	84 mg/kg
Radius	135pm
Melting point	1455°C
Boiling point	2913°C
Configuration	(Ar) 3d8 4s2
Discovered	1751 – F. Cronstedt

Father Christmas joins famous scientists on the periodic table. Nickel is named after one of its reddish ores, which German miners called *'kupfernickel'*, or Saint Nicholas' copper.

Nickel is a hardy and corrosive-resistant metal used to plate steel and iron. When it is used to plate jewellery, some people find themselves allergic to the tiny quantities that can dissolve in their sweat. Nickel coats the inside of cans to prevent food reacting with the tin, although in cans of peas black spots can be seen as it reacts with sulfur released by some bacteria to form nickel sulphide. Although the US 5¢ piece is fondly called the nickel it actually only contains 25% of the metal.

Cheap extraction
When passing noxious carbon monoxide gas through nickel valves, Sir Ludwig Mond found that the valves repeatedly failed and leaked. What he discovered in 1890 was that, remarkably, the nickel was reacting with the gas to form nickel carbonyl. Unlike those made by other metals, this compound had an extremely low boiling temperature of just 42°C. Carbonyls are thermally unstable and nickel carbonyl begins decomposing back into nickel metal and carbon monoxide at around 180°C. On stumbling across a failure in the factory valves, the German-born Mond had found a very simple and cheap way to extract nickel. He established the company now known as ICO and made himself a very rich man.

Starting out in life
Nickel was used by ancient life to harvest energy from the carbon monoxide-rich early atmosphere. It remains central to a number of enzymes that form essential parts of the carbon cycle. They convert carbon monoxide into carbon dioxide, carbon dioxide into acetate, and acetate into methane where it is released back into the atmosphere.

Nickel takes its name from the ore Saint Nicholas' copper, *kupfernickel* in German.

Copper
Relatively reactive redhead

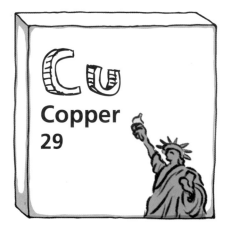

Atomic number	29
Atomic weight	63.546
Abundance	60 mg/kg
Radius	135pm
Melting point	1085°C
Boiling point	2562°C
Configuration	(Ar) 3d10 4s1
Discovered	9000 BCE

Copper was essential in the progress from the Stone Age to the Bronze Age, as to create bronze, two parts of copper are needed for every one part of tin. Bronze has appeared in archaeological sites around the globe, showing that copper has been mined and used by man for over 10,000 years.

The name for the metal comes from the Roman name for the island of Cyprus, *Cuprum*, where the Romans mined the majority of copper for their empire.

Colours in compound connections

The reddish colour of copper comes from electrons moving between the half-full 3d and half-full 4s subshell. The shiny surface of copper slowly blackens when exposed to air, forming copper oxide. With extended exposure copper structures like the Statue of Liberty develop a layer of verdigris (copper (II) carbonate).

Electricity and political friction

Copper is happy to release free electrons to swim amongst ions within the metal. This makes it

Carbon dioxide in the air reacts with copper over time to form the distinctive green of verdigris (copper (II) carbonate).

an excellent conductor of heat and electricity, which requires the easy movement of electrons through a material. Copper is very ductile, which means it can be stretched into very thin wires. For these reasons copper is used widely to connect our modern world, supplying electricity and data to our doorsteps: there is over 150kg of copper associated with each person in Europe and the US. As China, India and other developing countries rapidly build a similar infrastructure, the price of copper has rocketed; it costs four times more today than it did just five years ago. However, there is not enough copper in the known global reserves for all countries in the world to have the same level of infrastructure enjoyed by Europe and the US.

Giving you that buzz

In small quantities, copper plays a key role in life for every species on Earth. Most processes rely on the metal's ability for single electron transfer, switching between its $Cu+$ and Cu_2+ ions. Copper-containing enzymes use this property in the process of respiration to release energy from glucose. Another enzyme, tyrosinase, converts the messenger hormone chemical tyrosine into L-dopa which is a precursor to the fight-or-flight hormone adrenaline. L-dopa is also used as a treatment for Parkinson's disease as it is broken down by the body to form dopamine, which mediates otherwise broken communication links between brain nerve cells.

Blue blood

While the majority of life uses iron molecules to transport oxygen to cells, some organisms use copper instead. The bluish molecule haemocyanin contains two atoms of copper and effectively binds to a molecule of oxygen (O_2). Unlike haemoglobin the haemocyanin molecule is not found within specialised blood cells but instead within the blood itself. The crabs, lobsters, octopi and other invertebrates that rely on this system of oxygen delivery therefore have bluish rather than red blood. Land organisms which also employ the same molecule are spiders such as the tarantula, the emperor scorpion and a number of centipedes.

Tarantulas pass oxygen around their body using blue copper-based blood.

Zinc

Unappreciated and understated

Zn

Zinc
30

Atomic number	30
Atomic weight	65.38
Abundance	70 mg/kg
Radius	135pm
Melting point	429°C
Boiling point	907°C
Configuration	(Ar) 3d10 4s2
Discovered	1000 BCE

At the end of the transition metals of period 4, zinc likes to hold on to the electrons in its full 4s and 3d subshells. Strictly speaking, none of group 12 are really transition elements; they are usually referred to as 'honorary' transition metals.

Wanting to hold onto its electrons results in a lack of reactivity that makes the metal perfect for coating iron and steel to prevent them from rusting; it is used to galvanise everything from garden gates to corrugated roofing. The most attractive use of zinc metal has to be in forming an alloy with copper to make brass. Shiny and lustrous, brass has been used extensively since Roman times for decoration, musical instruments and memorials. Zinc metal is also used in zinc-carbon and zinc-alkaline batteries as the cathode.

Zinc provided strength to Roman swords as well as decorating their helmets and armour.

Sunblock and smelling good

Zinc oxide is great at absorbing ultra-violet light and is used in sunscreen to protect us from those harmful rays. A non-toxic white powder, it also finds widespread use in paints and mineral-based make-up. Zinc also helps us on a day-to-day basis in the form of zinc pyrithione in anti-dandruff shampoo and zinc chloride in deodorants. Where zinc really gets exciting, though, is in the field of organic chemistry.

A new field of science

Edward Frankland, of valence theory fame (see Mendeleev and the Modern Table), heated ethyl iodide (C_2H_{5l}) with zinc powder in a sealed glass tube. He was hoping to free ethyl radicals, C_2H_5+, but instead got the shock of his life. On adding a drop of water to the substance produced, a blue-green flame shot out over a metre from the tube. The explosive reaction of a substance with water or air earns it the name pyrophoric, from the Greek 'pyrophoros', meaning fire-bearing. Unintentionally, Frankland had given birth to a new field of science: organometallic chemistry.

The experiment created the first organometallic compound, diethylzinc, $(C_2H_5)^2Zn$, in which two ethyl groups bond with a zinc atom. This area of science bridges the gap between organic, carbon-based, and inorganic chemistry. It produces catalysts which speed up industrial reactions that produce the world's plastics. The compounds are also essential in making semiconductor electronics such as the humble light-emitting diode (LED) found in TV screens and other electronics across the world.

Breaking and building DNA

While the other members of group 12 are highly toxic to life, zinc plays a central role in its continuation. The metal is found in small protein structures called zinc fingers, with one or more zinc ions at an active centre. These ions provide stability to unwieldy strand like molecules, folding them at the perfect angle to break or assisting them to form new bonds. DNA and its cousin RNA are just two examples of such unwieldy molecules. Without these zinc fingers, the unzipping, copying and reassembling of DNA would not be so efficient.

Zinc may not have the colourful range of chemistry shown by the transition metals of period 4 but it is far from boring. It modestly works away in the background ensuring the continuation of life and rust-free steel, without ever trying to attract attention.

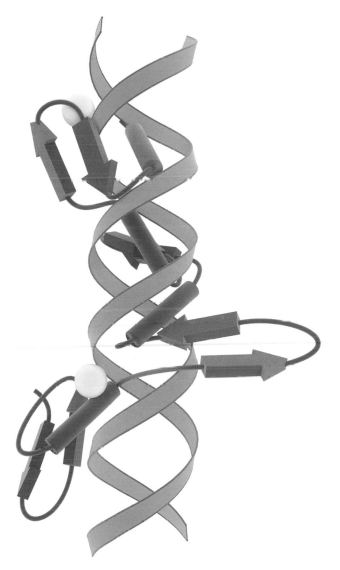

Zinc fingers pry apart a helix of DNA so that copies of the proteins it encodes can be made.

Yttrium

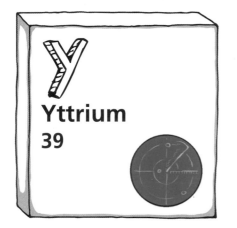

Y

Yttrium
39

Atomic number	39
Atomic weight	88.9058
Abundance	33 mg/kg
Radius	180pm
Melting point	1526°C
Boiling point	3336°C
Configuration	(Kr) 4d1 5s2
Discovered	1794 – J. Gadolin

Although it has not found much interest among chemists, yttrium is a rising star in the world of physics. Yttrium is one of four elements that owe their names to the small Stockholm suburb of Ytterby. The Apollo missions showed that it was far more abundant on the Moon than here on Earth.

High temperature but still cold

Until the mid-1980s no one had really cared about yttrium – that was until the discovery of 'high temperature' superconductors. Cool a metal with expensive liquid helium to near absolute zero (-273°C) and it no longer resists the movement of electrons flowing through it. Such zero electrical resistance materials are called superconductors. In the mid-1980s the compound yttrium barium copper oxide was observed to superconduct at the much higher temperature of -178°C. This is below the boiling point of inexpensive liquid nitrogen, making it a much cheaper superconductor to use.

Ceramic wires ?

The compound is a ceramic that is difficult to make into wires or films, which means it finds little current use. But it is hoped that one day this yttrium compound, or one similar, can produce affordable MRI scanners and other diagnostic machines.

Gems and RADAR

Yttrium aluminium garnets (YAG), $Y_3Al_2(AlO_4)$, are used as fake gemstones because of their high refractive index (see Zirconium for more).

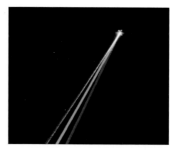

Yttrium plays an essential part in lasers, which find use in many areas of our modern lives.

Add a dash of a lanthanide element, and these clear stones take on a whole new life. Such crystals sit at the heart of semiconductor lasers, which emit a wide range of light. They are also used as microwave filters, essential to the operation of radio detection and ranging (RADAR).

Zirconium

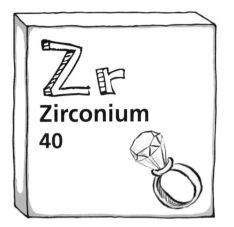

Zr
Zirconium
40

Atomic number	40
Atomic weight	91.224
Abundance	165 mg/kg
Radius	155pm
Melting point	1855°C
Boiling point	4409°C
Configuration	(Kr) 4d2 5s2
Discovered	1789 – H. Klaproth

Yellow-coloured gemstones called zircon have been used in jewellery for thousands of years, and give the element its name. These stones sparkle and shine because of the main ingredient, zirconium oxide, while other elements provide the colour.

Sparkle

Today zirconium oxide is used to fabricate cubic zirconia gems, which sparkle with more brilliance than diamonds. The sparkle arises from a high value of a property known as refractive index. When light enters the material, a lot of it becomes trapped and bounces around inside the gem. When the light eventually escapes from the face of the gem, it is released over a limited range of directions. This gives a sparkle when the stone is looked at from different angles. Zirconium oxide is extracted from abundant zircon; there are whole beaches in Australia made from zircon sand.

Nuclear holders

Zirconium metal turns out to be the perfect material to use inside nuclear reactors, where tubes of the metal contain the nuclear fuel. Even in this environment, zirconium does not become radioactive because neutrons pass right through it.

It does not corrode at low temperatures, but like most metals it will eventually react with steam if

Zirconium containers of nuclear fuel can react with steam at the extremely high temperatures resulting from a nuclear meltdown. This produces hydrogen gas, which leads to explosions; it is not the nuclear fuel that explodes.

hot enough. When this happens the zirconium strips the oxygen from the gaseous water, leaving behind hydrogen gas. Explosions at power stations like Chernobyl, and more recently Fukushima, were not caused by the nuclear fuel but instead through ignition of this flammable gas, which had accumulated from this reaction.

Niobium

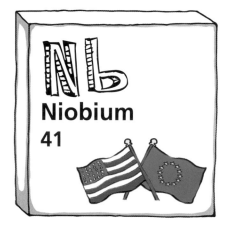

Atomic number	41
Atomic weight	92.9064
Abundance	20 mg/kg
Radius	145pm
Melting point	2477°C
Boiling point	4744°C
Configuration	(Kr) 4d4 5s1
Discovered	1801– C. Hatchett

The start of the 19th century saw scientists around the globe in a race to discover new elements. This inevitably led to international rivalries, not to mention debate over the naming of new discoveries.

Columbium

In 1801 the Englishman Charles Hatchett experimented with a mineral called columbite, from Massachusetts, USA via the British Museum. He produced what he thought to be a new metal and named it 'columbium' after the mineral. His result was met with criticism by fellow English chemist William Hyde Wollaston. After conducting the very same experiment, Wollaston concluded that the solid was just a compound of recently discovered tantalum. Soon after this Hatchett moved away from science to start a successful business building horse-drawn carriages.

Niobium

It was not until 1844 that the German chemist Heinrich Rose proved Hatchett right all along. The precipitate made was a mix of both tantalum and a new metal oxide. With Hatchett out of science, Rose chose a new name for the element, niobium. The name comes from Niobe, daughter of the mythological Greek underworld figure Tantalus, who lends his name to tantalum.

Debate and the rise of the IUPAC

By this time, however, 'columbium' had been printed in a number of chemistry textbooks. This name was preferred by American scientists, and tension grew between them and their European counterparts. The raging debate, and others like it, led to the creation of the International Union of Physical and Applied Chemistry (IUPAC) in 1919. The new governing body brokered a deal in which US scientists agreed to call element 41 niobium if element 74, which Europeans called wolfram, be named tungsten. The IUPAC is still tasked today with deciding naming rights for new elements.

Arguments raged regarding the naming rights of the elements.

Molybdenum

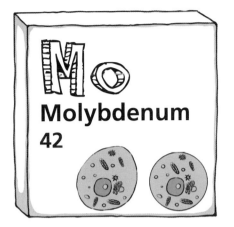

Mo
Molybdenum
42

Atomic number	42
Atomic weight	95.95
Abundance	1.2 mg/kg
Radius	145pm
Melting point	2623°C
Boiling point	4639°C
Configuration	(Kr) 4d5 5s1
Discovered	1781– P.J. Helm

If you are a Douglas Adams fan you will know that 42 is the answer to the ultimate question: What is the meaning of life, the universe and everything? Element 42 certainly plays a major part, perhaps not in the meaning, but certainly in the creation of life as we know it.

Building blocks of life

Limited amounts of molybdenum held back the evolution of multicellular life in the oceans for around two billion years. It was not until oxygen in the Earth's atmosphere rose to significant levels that soluble MoO_{42} ions were formed from its oxide salts. As soon as this happened the ocean-bound bacteria flourished and created enough usable nitrogen compounds for other organisms to grow.

Ions of the metal are core to essential nitrogenase enzymes, which fix nitrogen gas to form nitrogen compounds. The enzymes are found in single-cell bacteria but not in the cells of higher life forms. Nitrogen compounds are essential to make nucleic acid building blocks of DNA and other similar molecules (see Nitrogen). Without bacteria to make these compounds from atmospheric nitrogen, higher forms of life could not multiply and grow.

Not lead

Molybdenum owes its strange name, in a convoluted way from Greek, to a common mistaken identity of its ores with that of lead. Graphite was another false guise for molybdenum which, despite many 'discoveries', was not truly recognised as an element until 1778.

Without this molybdenum-containing enzyme converting nitrogen (N_2) from the air into ammonia (NH_3) then we could not have the nucleic acids which encode life.

Technetium
Elementary, dear Segrè

Technetium
Tc
43

Atomic number	43
Atomic weight	(98)
Abundance	3x10-9 mg/kg
Radius	135pm
Melting point	2157°C
Boiling point	4265°C
Configuration	(Kr) 4d5 5s2
Discovered	1937 – C. Perrier and E. Segrè

Sir Arthur Conan Doyle's super-sleuth Sherlock Holmes once said, "Detection is, or ought to be, an exact science." The case of element 43 was, however, one far from elementary and it was the last of the four 'missing' elements originally predicted by Mendeleev to be found.

Searching

After the discovery of scandium, germanium and gallium, the hunt for eka-manganese intensified. Despite many false sightings, it eluded scientists for well over a century. It was not until 1937, almost 160 years after its existence had been predicted, that Carlo Perrier and Emilio Segrè at the University of Palermo solved the case.

The 60-inch cyclotron particle accelerator, built by Ernest Lawrence and colleagues at Berkley, which was involved in the discovery of technetium and heavier radioactive elements.

Light but unstable

The problem was that element 43 is the lightest unstable element in the periodic table. There are no isotopes of this element with a stable nucleus; all decay within a relatively short period of time. Half of its most stable isotope ^{98}Tc, decays every 4.2 million years, a property known as its half life. While this sounds like a very long time, the age of the Earth is over 1000 times greater at around 4.5 billion years old. New atoms of this element are only produced in tiny quantities from the decay of heavier uranium. Any trace of this element on a brand new Earth would have long since decayed.

Atom smasher

In 1936 Segrè had visited Ernest Lawrence's atom-smashing cyclotron particle accelerator in Berkley California, USA (see Atom Smashers). Segrè had an idea that new elements might be created in the energetic collisions in the machine. He requested, and Lawrence sent him, a sample of deflector foil made of lighter element 42, molybdenum. Segrè was a particle physicist by trade and so he teamed up with mineralogist Perrier for help with identifying the elements present. After painstaking work they managed to isolate two isotopes of element 43. As these atoms were man-made in a machine, the pair gave the element the name technetium from the Greek for artificial: technitos. It is produced today in largest amounts in decay products of radioactive waste from nuclear reactors.

Seeing inside

The chemistry of technetium is very similar to rhenium or manganese above and below it in the periodic table. It can therefore form compounds with an array of different elements. Coupled with its radioactivity, it finds varied use as a medical diagnostic tool. The isotope technetium-^{99}m has a short half life of just six hours and releases a very distinct energy of gamma radiation. Carefully selecting which element bonds to a technetium atom determines where in the human body it will be absorbed. Technetium can therefore be directed to image and diagnose a host of different medical conditions. This elusive metal has gone from being the subject of investigation to becoming one of medicine's detectives.

If inside the body, the short-lived 99mTc isotope will emit gamma radiation, which passes through the soft tissue to be detected by cameras, giving a detailed internal picture.

Ruthenium

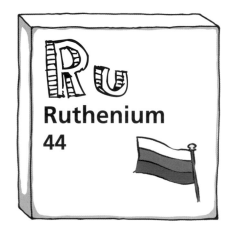

Ru
Ruthenium
44

Atomic number	44
Atomic weight	101.07
Abundance	0.001 mg/kg
Radius	130pm
Melting point	2334°C
Boiling point	4150°C
Configuration	(Kr) 4d7 5s1
Discovered	1844 – K. Claus

This element owes its name to the Latin name, Ruthenia, for what is now modern Russia. It was discovered in 1844 amongst platinum ores by Karl Karlovich Klaus working in Kazan, Russia. It is extremely rare and similar to neighbouring metals, although it produces rather stable compounds.

Sunlight

Tris(bipyridine)ruthenium(II) chloride, with a much catchier nickname of Ru-bpy (roo-bi-pee), is extremely photo-stable. It is able to absorb a large spectrum of ultraviolet and visible light without breaking down into simpler molecules. There is extreme research interest into its use for harvesting solar energy.

Ruthenium catalysts are essential for creating pharmaceuticals.

Goldilocks...

In the 1960s the field of organometallic chemistry was flourishing (see Zinc). Lots of chemists were boiling up metals with organic compounds and investigating the behaviour of whatever was made. It turns out that ruthenium forms 'goldilocks' organometallic compounds, with just the right balance between stability and reactivity to be really useful.

... and bond swapping

Life is comprised of molecules with carbon-bonded backbones, and the breaking and making of carbon bonds form new organic materials. This can be done naturally or artificially, using catalysts. In 1992 American chemist Bob Grubbs discovered a very important organometallic ruthenium catalyst which mediated a process called metathesis, which means 'changes places'.

The catalyst breaks and makes double bonds between carbon atoms, causing groups of atoms to change places. Other platinum group catalysts can perform the same task, but Grubbs' ruthenium catalyst is the only one stable enough to use in air. Without it we could not produce and supply life-saving drugs at the scale necessary.

Rhodium

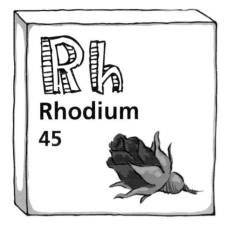

Rh
Rhodium
45

Atomic number	45
Atomic weight	102.9055
Abundance	0.001 mg/kg
Radius	135pm
Melting point	1964°C
Boiling point	3695°C
Configuration	(Kr) 4d8 5s1
Discovered	1804 – H. Wollaston

At the heart of the transition block there are six precious metals with similar physical and chemical properties: ruthenium, rhodium, palladium, osmium, iridium and platinum. This collection of metals is called the platinum group, after the first of them to be discovered.

Hardy but helpful

These metals are chemically resistant and hardy, noble metals. They are also brilliant catalysts, used to speed up a variety of chemical processes. All of the group are rare in nature and are often found together in ores because they form similar compounds. This also makes separating them from each other a difficult task. In fact rhodium was discovered by William Hyde Wollaston as an impurity only after he had started selling his newly discovered palladium. He named it rhodium from the Greek *rhodon*, meaning rose-coloured, as it produced vivid red salts.

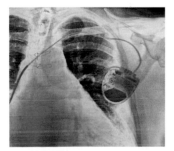

Coloured chest X-ray showing a pacemaker, which supplies electrical impulses to the heart through a rhodium-platinum wire.

Medical specialist

Pure rhodium foil filters X-rays which are emitted by radioactive sources, and are used for cancer, particularly breast cancer, diagnosis. Because it is the rarest of all non-radioactive metals, however, rhodium finds little other use as an elemental metal. It is predominantly alloyed with platinum. Platinum wire with a little rhodium added forms very stable thermocouples, able to measure temperatures up to 1800°C. The corrosion resistance of the wire allows it to survive the hostile environment experienced by pacemaker implants. Rhodium-platinum wires relay the electrical impulses direct to the heart muscle to keep it beating.

Protector

The catalyst ability of platinum-rhodium alloys is used inside catalytic converters of automobile exhausts. These metals catalyse the breakup of nitrogen oxides into nitrogen gas and oxygen. If nitrogen oxides enter the atmosphere they dissolve in water to form acids, a source of acid rain.

Palladium

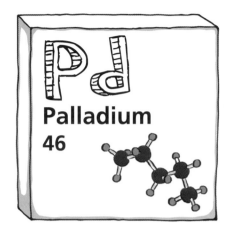

Pd

Palladium

46

Atomic number	46
Atomic weight	106.42
Abundance	0.015 mg/kg
Radius	140pm
Melting point	1555°C
Boiling point	2963°C
Configuration	(Kr) 4d10
Discovered	1803 – W.H. Wollaston

Roll up, roll up and buy some of this fantastic new metal! Englishman William Hyde Wollaston was an industrious chemist and physicist. In 1802, while carefully precipitating metals from processed ore, he discovered a new metal. Using his science to make money, he quickly set himself up in business selling it.

Fashionable name

To boost sales, Wollaston named the metal after a recently discovered 'planet' which had made big news just two years earlier. This celestial body was named 'Pallas', following the tradition of naming planets after iconic symbols. It was not until years later that it was understood not to be a planet, but the second-ever asteroid to have been discovered. His advertising trick worked, however, and orders came in from far and wide for this 'new silver'. Sales were rocked slightly when one customer, after a number of chemical tests, argued that it was not a new metal. This forced Wollaston to eventually write a paper explaining his method of obtaining it.

Investigating induction

Wollaston was a fantastic all-round scientist, pioneering work in electricity, optics and even biology. One accidental discovery of his was that moving magnetic fields induce an electric

An induction coil made by Michael Faraday. He used Wollaston's palladium and other platinum group metals in his landmark series of experiments on electricity and magnetism.

current in nearby metal wires: a process known as electromagnetic induction. When Wollaston died he bequeathed samples of rare palladium and platinum to the Royal Society. These were used by Michael Faraday in his landmark experiments with electricity and magnetism. Despite the essential part these metals played, Faraday ignored Wollaston's accidental discovery of electromagnetic induction when writing his own account of the phenomenon some ten years later. Today palladium finds most use as a catalyst in the Heck reaction, which connects carbon-carbon bonds in organic chemistry.

Silver

Ag

Silver

47

Atomic number	47
Atomic weight	107.8682
Abundance	0.075 mg/kg
Radius	160pm
Melting point	962°C
Boiling point	2162°C
Configuration	(Kr) 4d10 5s1
Discovered	5000 BCE

Silver is not very reactive and is one of only a few metals found in elemental form in nature. It has been known of and used since ancient times thanks to people tripping over or digging up lumps of it. The metal's low reactivity comes from a lack of interest in sharing or exchanging electrons thanks to a full d subshell.

Wartime lending

The same d electrons become delocalised in pure silver and make it a brilliant conductor of both heat and electricity. During World War II thousands of tonnes of silver were borrowed from the US treasury by scientists working on the Manhattan Project. It was stretched to make wire, then spun into coils, to make strong electromagnets essential to enrich uranium for nuclear weapons.

Capturing a memory

Silver compounds have been used for centuries in film photography. These compounds are extremely photosensitive, breaking down in the presence of light to form small deposits of silver metal. Deposited silver then acts as a catalyst for chemical reactions within developing fluids and leads to a fixed image. Heating these compounds leads to the same breakdown and is used to form a thin layer of the metal upon glass to make mirrored surfaces.

Rainmaker

A form of silver iodide, found in the mineral idogaryite, has a crystalline structure similar to water ice. Each year 50,000 kg are dropped by aeroplanes into clouds where they act as starting points for water vapour to freeze. This ice eventually thaws and falls as rain over farmland desperate for water.

The metal has not been shown to be toxic to humans and is even used for colouring food products as additive E174 in the EU.

Many everyday items contain silver, which effectively kills bacteria with which it comes into contact.

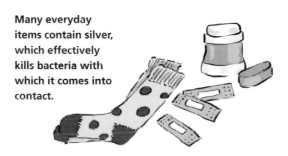

Cadmium

Cd

Cadmium
48

Atomic number	48
Atomic weight	112.414
Abundance	0.159 mg/kg
Radius	155pm
Melting point	321°C
Boiling point	767°C
Configuration	(Kr) 4d10 5s2
Discovered	1817 – Hermann, Stromeyer & Roloff

Cadmium's chemistry is very similar to that of its heavier brother mercury, and both are extremely toxic.

It hurts, it hurts

Cadmium drains calcium deposits from bones, leaving them very porous and brittle. Around 1912, residents of the Toyama Prefecture in Japan began to suffer from these effects after eating rice grown in fields rich in cadmium salts. The locals named the excruciating pain they felt in the spine and joints 'itai-itai' disease, which literally translates as 'it hurts, it hurts'. Such cadmium poisoning is rare, though, because levels must be very high to overwhelm our body's effective defences.

Warm colouring

Bright orange cadmium sulphide colours plastics or glazes, securely bonded into the material. Underground gas pipes are often coloured with this compound, and the cadmium also adds resistance to weathering. Artists have also used cadmium salts for years to make the most vivid and stable red, yellow and orange colours.

Quantum computing

Cadmium semiconducting compounds are leading the way into a new world of technology. Cadmium

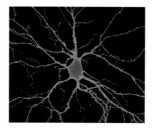

The blue-ultraviolet cadmium laser excites atoms in a fluorescent protein to light up a neuron taken from the brain of a mouse.

selenide and cadmium sulphide are finding use in the first generation of quantum dot computers. These materials trap photons, particles of light, which are then manipulated by powerful lasers to process information. Cadmium can also be used to make the lasers. Common helium cadmium lasers produce intense blue-ultraviolet light used by fluorescence microscopes.

Dialling down a nuclear power station

Cadmium metal finds another use not from the chemistry of its electrons but the behaviour of its nucleus. In the presence of silver nuclei, a cadmium nucleus is very effective at absorbing neutrons. This is essential to control the otherwise runaway decay of nuclear fuel.

Hafnium

Atomic number	72
Atomic weight	178.49
Abundance	3 mg/kg
Radius	155pm
Melting point	2233°C
Boiling point	4603°C
Configuration	(Xe) 4f14 5d2 6s2
Discovered	1923 – D. Coster & C. de Hevesy

Early periodic tables sorted the elements in order of their atomic weight, as a measure of the mass of all protons, electrons and neutrons within an atom. In his second table, however, Mendeleev chose to switch the ordering, to better group together elements with similar physical and chemical properties.

Mendeleev placed cobalt ahead of nickel despite their having almost the same atomic weight, and tellurium ahead of iodine despite its having a larger atomic weight. This hinted that there may be some undiscovered underlying law which truly ordered the elements, rather than atomic weight.

New elements expected

In 1913, Englishman Henry Moseley noticed a relationship between the wavelength of X-rays absorbed by, and the position of, each element on the table. The link led Moseley to predict that there existed missing elements, at atomic numbers 43, 61, 72 and 75, and provided evidence that there are no other gaps from aluminium to gold.

Element 72 takes its name from the Latin for Copenhagen, Hafnia, where the element was first isolated in 1922 by George Charles de Hevesy and Dirk Coster. It is as abundant as other rare elements but difficult to isolate from the very similar atomic weighted zirconium.

Extreme essential

Expensive hafnium is used for its ability to absorb neutrons; hafnium rods control the radioactive decay chain within extreme nuclear reactors. They are used in place of weaker cadmium in high water pressure reactors found on submarines, because of the metal's surprising resistance to corrosion. A mixture of carbon, tungsten and hafnium (tungsten-hafnium carbide) has the highest melting point of any known compound at 4125°C.

A number of elements were discovered in Copenhagen, but it is Hafnium that owes its name to the city.

Tantalum

Tantalum
73

Atomic number	73
Atomic weight	180.9479
Abundance	2 mg/kg
Radius	145pm
Melting point	3017°C
Boiling point	5458°C
Configuration	(Xe) 4f14 5d3 6s2
Discovered	1802 – G. Ekeberg

King Tantalus, in Greek mythology, was punished by the gods after stealing their secrets. He was placed in a pool of water, below a tree with low-hanging fruit. Every time he reached up for the fruit or bent down to drink the water they receded away from him.

Stable when excited

Swedish chemist Anders Ekeberg was reminded of this story when the new metal he found in 1802 refused to react with acids, and he named the metal tantalum. Tantalum is also the only element to have an inert radioactive isomer; tantalum-180 should be a radioactive isotope but when it is in an excited state is perfectly stable. It is as if by adding energy to these atoms there is some shelf they can sit on and not fall back down to lose energy through decay.

Tantalum atoms fluorescing under ultraviolet light in the tantalite mineral.

40 milligrams, of this rare metal in each device and around 1.8 million kilograms make it into electronics each year.

Transistor, tantalum in disguise

The mobile phones we carry around are the finest example of how industry has driven the miniaturisation of electronics. This has been possible through the use of tantalum to make ever-smaller electronic components called capacitors. Without this technology we would be stuck with brick-sized phones from the 1990s rather than the full computing power of today's smartphones. There is a tiny amount, around

Politically fraught

A large proportion of the world's supply of tantalum comes from rich deposits of the ore columbite-tantalite, found in large amounts in The Democratic Republic of the Congo (DRC). Profits from mining this ore, and other rare earth metals, have helped fund both sides of the DRC's long-running civil war. Human rights groups and the UN have condemned the trade of the ore from the DRC.

The secret life of the periodic table

Tungsten

Tungsten
74

Atomic number	74
Atomic weight	183.84
Abundance	1.3 mg/kg
Radius	135pm
Melting point	3422°C
Boiling point	5555°C
Configuration	(Xe) 4f14 5d4 6s2
Discovered	1783 – J. José & F. Elhuyar

There is controversy surrounding the name of element 74.

In 1781 Swedish chemist Carl Wilhelm Scheele noted that a new acid, containing a new metal, could be formed from a mineral named 'tung sten', which translates as 'heavy stone'. Two years later the Spanish Elhuyar brothers, Juan and Fausto, isolated the metal from an acid they had formed from the ore wolframite. The Elhuyar brothers thus named their metal wolfram, but English-speaking chemists had already adopted the name tungsten.

While both names were recognised by the IUPAC, tungsten was preferred, and in 2005 it was chosen as the international name for element 74. The argument was that the IUPAC is an English-speaking society and so should adopt the name used among English speakers. The symbol of the element, however, W for wolfram, still bears the history of this debate.

Hottest element around

Tungsten is tough and has the highest boiling point of any element at 5930°C. Flow of electricity is naturally resisted by all metal wires, which causes them to heat. As an object heats up it glows first red, then orange, then yellow, and if hot enough white. Tungsten wire is used in traditional filament light bulbs as it can reach white hot. Surrounded by a halogen gas the tungsten can be run hotter, producing a more intense light. Such bulbs are used in high-end car headlights.

Hard and colourful

Tungsten carbide, where the metal is mixed with carbon atoms, is extremely tough and commonly used on the tip of ball-point pens. Tungsten oxide is also one of only a handful of compounds that are electrochromic, changing colour in the presence of an electric field. These compounds are used to make colour-changing smart glass and displays.

With a density similar to that of gold but much cheaper to purchase, there have been cases where gangs have tried to sell counterfeit gold bars: simply tungsten plated with gold.

Rhenium

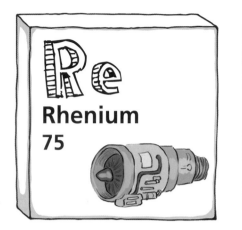

Rhenium

75

Atomic number	75
Atomic weight	186.207
Abundance	7x10-4 mg/kg
Radius	135pm
Melting point	3186°C
Boiling point	5596°C
Configuration	(Xe) 4f14 5d5 6s2
Discovered	1925 – Noddack, Tacke & Berg

This was the last stable element to be officially discovered, in Germany in 1925, by Walter Noddack, Ida Tacke and Otto Berg. The trio processed around 660kg of the ore molybdenite and ended up with just 1 gram of the metal, which they named after the nearby river Rhine. Today it is extracted much more efficiently in the refining of molybdenum and copper, but remains rare and expensive.

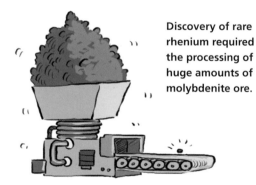

Discovery of rare rhenium required the processing of huge amounts of molybdenite ore.

Discovery and misidentification

In 1908 Japanese chemist Masataka Ogawa was widely discredited for claiming to have discovered element 43, now named technetium. However, research published in 2004 showed that he had in fact isolated rhenium. To date the only officially recognised discovery of an element from Japan, or indeed the continent of Asia, is element 113 (see Hall of Fame).

All over the states

Lying in the middle of the transition block, rhenium can take on one of the widest range of oxidation states from -3 through to +7. This ability affords rhenium a role as a brilliant catalyst, for example in converting natural gas into high-octane automobile fuel. Other uses being investigated are in solar cells to capture light and for splitting water into hydrogen and oxygen to be used as fuel.

Single crystal jet engine

Rhenium metal adds strength when alloyed with nickel, and single crystals of the metal are grown to produce whole turbine blades for jet engines. They are turned by exploding fuel and used as they do not deform despite the intense temperature and mechanical stress.

The secret life of the periodic table

Osmium

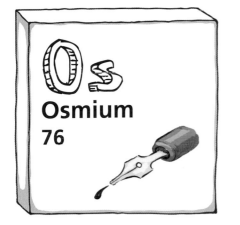

Os

Osmium

76

Atomic number	76
Atomic weight	190.23
Abundance	0.002 mg/kg
Radius	130pm
Melting point	3033°C
Boiling point	5012°C
Configuration	(Xe) 4f14 5d6 6s2
Discovered	1803 – S. Tennant

Osmium's fifteen minutes of fame was stolen by iridium, discovered alongside it in 1803 by the English chemist Smithson Tennant. He adored the rainbow sheen shown by iridium elements, but described the 'pungent and penetrating smell' of osmium as one of its 'most disgusting characters'.

Stench

These observations were reflected in naming the elements, with osmium named from *osme*, the Greek for odour. The smell emanating was from the stable but easily evaporated osmium tetroxide (OsO_4). An uncanny ability for this compound to bind to double carbon bonds gives it a use as a biological staining agent in specialist microscopy and fingerprint detection. Those who use it need to be very careful though; if it gets into the eye it can stain the retina, causing blindness.

Densest metal

In the 1990s, osmium got its time in the spotlight when detailed research showed it to be the densest known metal, stealing the crown from iridium. Due to its density and strength, the metal is used in various alloys with other platinum group metals, which can be found in the tips of fountain pens or the needles of early vinyl record players. Osmium's high boiling point saw it used for a brief period instead of tungsten filaments in bulbs. The name of the German lighting company Osram highlights this early use as it is a combination of the names osmium and wolfram (tungsten). Today, less than 100kg of osmium is refined and used annually.

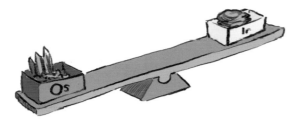

In the 1990s osmium took the title of densest transition metal from iridium.

Iridium
Rarer than rainbows

Ir
Iridium
77

Atomic number	77
Atomic weight	192.217
Abundance	0.001 mg/kg
Radius	135pm
Melting point	2446°C
Boiling point	4428°C
Configuration	(Xe) 4f14 5d7 6s2
Discovered	1803 – S. Tennant

Element 77 was discovered in 1803 by Englishman Smithson Tennant from insoluble impurities in platinum. The salt compounds it makes are varied and vivid in colour, like the iridescent wings of a dragonfly.

For this reason, Tennant named it iridium after the Greek goddess Iris, who personified the rainbow. As with all transition elements the colours arise from the multiple oxidation states the metal can take.

Brought back down to Earth

Iridium is the rarest element in the Earth's crust because, along with the other heavy metals, most of it sank into the iron core when the Earth was still molten and young. It is found in much higher quantities in extra-terrestrial asteroids and

Layers of iridium seen worldwide suggest that a massive meteorite hit the planet 66 million years ago, which could have been responsible for the extinction of the dinosaurs.

Iridium metal is tough enough to withstand the harsh conditions at the tip of spark plugs.

meteorites, so rich deposits of iridium are usually a clear sign of an asteroid impact site.

In the 1980s the team of Luis Alvarez, his son Walter, Helen Michel and Frank Asaro discovered a thin layer in sedimentary rock that showed a large concentration of the element. In Italy they found this layer to have 30 times greater, and on the Danish island of Zealand 160 times greater, than the natural abundance of iridium. This geological layer is known today as the Cretaceous-Paleogene boundary and defines a short time period in our planet's history around 66 million years ago. The group hypothesised that this layer had come from the vaporisation of a large meteorite.

Dinosaur disaster

Such an event would throw tonnes of dust and debris into the upper atmosphere, blocking out the light of the sun. Taking years for the dust to settle, many plants and animals dependent upon sunlight would die out, and any large animals such as dinosaurs would have certainly perished. This is known as the Alvarez hypothesis, after the father and son leading the investigation. The iridium signature provides some of the most compelling evidence that a meteorite impact triggered a mass extinction.

I will not weather

Around six tonnes of iridium metal is used each year as it is the most corrosion-resistant metal known. Because it is so expensive it is used in tiny quantities or alloyed with other metals. Only iridium can survive the mechanical and thermal shock experienced in the electric contact of spark plugs that ignite fuel in combustion engines.

If measuring distance, the last thing you would want is for your ruler to change shape between readings. The standard metre bar made in 1889, to which all rulers were compared until 1960, is defined by a metal bar sealed inside a vault in Paris. It is made from 90% platinum and 10% hard-wearing iridium to prevent any change from occurring.

With such a high melting point, iridium is used to make crucible pots that can withstand the temperature of molten silicon. As the silicon cools slowly in these crucibles it forms large crystals, essential to the electronics industry (see Silicon).

Probing the nucleus

Iridium atoms were the first to demonstrate something known as the Mössbauer effect; the nucleus absorbs and emits gamma radiation without loss of energy from recoil. This process allows chemists to determine the energy levels of protons and neutrons in the nucleus.

Platinum

Pt

Platinum

78

Atomic number	78
Atomic weight	195.084
Abundance	0.005 mg/kg
Radius	135pm
Melting point	1768°C
Boiling point	3825°C
Configuration	(Xe) 4f14 5d9 6s1
Discovered	1735 – A. de Ulloa

The Spanish conquistadors of the 16th century only had eyes for gold. To their annoyance, when panning they kept coming across a great deal of a white metal instead. This was thought to be unripe, and was thrown back into the rivers to mature into gold.

So unappealing was the white metal that they viewed it as inferior to silver, giving it the name '*platina*' (inferior silver). This slight gives element 78 its name. Little did they know this white metal was much rarer than gold and far more precious.

Aloof

Platinum is a highly stable metal, both chemically and physically. It is the metal most resistant to chemical attack, only dissolving in a concentrated mixture of nitric and hydrochloric acid. This led to use in medical implements and prosthetics, non-corrosive laboratory containers and electrical contacts.

It is hard with an extremely high melting point, making it notoriously difficult to shape and mould. Such was the difficulty in purifying and working the metal that King Louis XVI of France believed that platinum metal was only fit for kings.

Automobiles past, present and future

Platinum also has a fantastic ability to catalyse reactions and is used widely by industry. The most common use today is within catalytic converters of automobile exhausts, where toxic carbon monoxide is converted to carbon dioxide.

Moving to hydrogen-powered cars would also see platinum take a lead role, as it catalyses the splitting of water into oxygen and the desired hydrogen fuel. Today rarity and demand make platinum much more precious than gold.

Spanish conquistadors believed gold to be more precious than platinum.

Gold

Gold
79

Atomic number	79
Atomic weight	196.9666
Abundance	0.004 mg/kg
Radius	135pm
Melting point	1064°C
Boiling point	2856°C
Configuration	(Xe) 4f14 5d10 6s1
Discovered	6000 BCE

If you are very lucky you could trip over a lump of gold in a field. It is chemically unreactive, so does not form compounds, but is also poor at naturally alloying with other metals.

Large lumps

The largest nugget of gold ever found was the Welcome Stranger nugget, found in Victoria, Australia in 1869, weighing in at over 71kg. Finds of large nuggets have led to gold rushes around the world throughout history. It is difficult to know how much gold is mined each year as it is kept a secret by those who mine it. The largest single hoard is 30,000 tonnes, and sits in the US Federal Reserve Bank in New York, belonging not just to the US, but 19 different nations.

Soft pieces of eight

Gold is the most malleable of all transition metals and soft enough to cut with a steel knife. Pirates used to bite coins to check if they were gold; if their teeth left an indentation then it was the real thing. Gold can be pounded into wafer-thin sheets just tens of atoms thick to decorate everything from buildings to food. The delocalised electrons present in all metals mean that just a few atoms are needed for the metal to shine.

Einstein's theory of relativity explains the change in energy of the electron orbits within gold that lead to the metal absorbing blue light and appearing yellow.

Einstein's colour

Gold glistens yellow thanks to Einstein's theory of special relativity. The inner electrons of a gold atom are held so tight that they travel at speeds approaching the speed of light. Travelling this fast, relativity tells us, the electrons' perception of distance changes. As the energy of electron orbitals depends on distance to the nucleus it all results in a change in energy of the electron shells. The outer 5s and 6d gold electrons absorb more blue light than other metals because of the energy change. This makes the metal look yellow overall, itself a mix of the reflected red and green light.

Mercury
Lethal liquid

Hg
Mercury
80

Atomic number	80
Atomic weight	200.592
Abundance	0.085 mg/kg
Radius	150pm
Melting point	-39°C
Boiling point	357°C
Configuration	(Xe) 4f14 5d10 6s2
Discovered	2000 BCE

The only liquid metal at room temperature, element 80 has transfixed people for over three millennia. Known commonly as quicksilver, the ancient Greeks named it 'hydragyrun', literally meaning 'liquid silver'.

Splitting salt

Mercury is liquid as its atoms want little to do with each other because of their size and d shell full of electrons. Like all transition metals, though, these electrons delocalise forming a sea around central metal ions, and can be forced to move as an electric current. Mercury finds industrial use as electrodes in the electrolysis of regular table salt, sodium chloride (NaCl). The process forms sodium metal in amalgam with a mercury cathode and chlorine gas. Most of the sodium is then used to make caustic soda (sodium hydroxide – NaOH) on reacting with water, with a by-product of hydrogen gas (H_2).

Millimetres of mercury

As with other metals, mercury expands when heated, which made it useful in thermometers; as the temperature increased the mercury would

Mercury compounds used to treat materials were linked to madness among hat-makers.

expand and rise higher up the thermometer. Due to its high toxicity, however, these thermometers are being rapidly replaced with alternatives that use dyed alcohol instead.

The metal is still used as an international temperature standard, from which all other temperatures are gauged. Mercury's triple point (a particular temperature and pressure where it can exist as solid, liquid and gas at the same time) at a pressure 50 million times less than that of normal atmosphere and a temperature of -38.83440°C, is used to gauge temperatures read around the world.

Mercury and man

As alluring as it is, mercury is deadly. Around a gram of the stuff is enough to kill you. Apart from direct exposure, it usually gets into our body through organometallic compounds. Some aquatic bacteria obtain energy in a chain of chemical reactions which end in their production. The greatest sources of them are, however, man-made: from chemical processing, burning of fossil fuels (containing trace amounts of mercury) or mining.

Many of the molecules in these compounds can be dissolved in fats and, if eaten, can be stored in animals. The biggest danger to humans is from eating fish that have eaten other fish. Tiny amounts of mercury in bacteria or plants quickly builds up inside fish that eat large amounts to survive. Other fish, like tuna, eat a large number of these smaller fish and the level of mercury continues to rise. This process of bioaccumulation leaves some predatory fish with lethal levels of mercury in their bodies. A number of events involving industrial leaks of the metal have ended in many human deaths from eating these fish.

Madmen

Mercury compounds don't even have to be eaten: many are very effectively absorbed through contact with the skin. The use of mercury has plummeted, and it is no longer used as mercury(II) nitrate, for instance in treating felt and furs to make the finest hats. It poisoned many a milliner, inducing hallucinations and other brain-related illnesses, which is where we get the phrase 'as mad as a hatter'. Romans used mercury compounds in cosmetics, which had the common effect of eventually disfiguring the very face they were intended to make beautiful.

Mining once more

In 2012 the European Union called for all countries to switch to energy-saving compact fluorescent light bulbs and previously closed mines reopened to supply the new demand for the mercury. In these bulbs, an electric current excites mercury to form a vapour and emit ultraviolet light. Phosphor coating the inside of the bulb absorbs the ultraviolet and re-emits visible light. Because of the mercury in each, many countries class used bulbs as hazardous waste.

Mercury vapour is excited by electricity inside energy-saving compact fluorescent light bulbs as it provides a natural spectrum of light.

Post-transition Metals
Bridging the gap

These metals are a collection that spans groups 13–15, forming a triangle which highlights the diagonal trends in the periodic table arising from a combination of electron configuration and atomic size.

Softies

Sandwiched between the transition metals and the metalloids, the post-transition metals show similar characteristics. They are mostly soft with poor strength and are often alloyed with transition metals to improve these mechanical properties. Aluminium, the lightest of the bunch, has typically low strength, but does have a high strength-to-weight ratio, making it useful in its elemental form.

Reaction

Post-transition metals are generally unreactive and lend their chemical inertness to other metals, through alloys or coating. This inert behaviour is due to the fact they are too far from noble gas status, so they would have to lose or gain a lot of electrons to become ideally stable. It also means that individual atoms tend to like isolation, giving these elements melting points below those of transition metals.

Aluminium shows a higher reactivity than the others because it does not have any d shell electrons. This brings it just three leaps away from obtaining a noble gas electron configuration, if only it could share electrons.

Covalent not conductor

Close to the metal/non-metal border, crystalline structures of these metals often show sharing of electrons between atoms through covalent bonding. This is very different to the ionic delocalisation of electrons seen in the transition metals, and leads to generally poorer conduction of electricity.

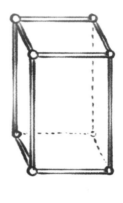

The post-transition metals behave increasingly less like metals and more like non-metals, forming different allotrope structures such as the metallic white tin (top) and the grey non-metallic tin (bottom).

The secret life of the periodic table

Aluminium

Atomic number	13
Atomic weight	26.9815
Abundance	82300 mg/kg
Radius	125pm
Melting point	660°C
Boiling point	2519°C
Configuration	(Ne) 3s2 3p1
Discovered	1925 – H.C. Oersted

It's everywhere, the most abundant metal in the Earth's crust, but it took until 1825 before pure aluminium was first extracted.

Tearing it free

Aluminium is found mainly in silicate mud, extremely well bound to oxygen atoms. It was not until the process of electrolysis was perfected (see Potassium) that electricity could wrench the metal from the compound. This was achieved in 1825 by the Danish chemist Hans Christian Oersted who commented unenthusiastically that 'it forms a lump of metal that resembles tin in colour and sheen'. In the mid-19th century the difficulty of obtaining the metal meant it was viewed as very valuable indeed. An aluminium bar held pride of place beside the crown jewels in the 1855 Paris Exposition and Emperor Napoleon III of France dined off aluminium plates with aluminium cutlery while less important guests ate off gold.

American English

Cornishman Humphry Davy named the element aluminum, after the source compound alum. However, shortly after this the International Union of Applied and Physical Chemistry standardised the suffix of the name to the more conventional '-ium'. In 1925 the American Chemical Society resurrected the original spelling and so, ironically, Americans use the name Englishman Davy intended.

Protect and serve

The metal does not react with air because of a nanoscopic protective coating of aluminium oxide. This means it finds use everywhere in our modern world, from TV aerials to food packaging. Scrunch up wrapping foil and you'll notice it is a soft metal, but alloyed with other metals aluminium lends its lightness to another's strength. It reduces the weight of materials so that massive planes can fly, and vehicles can go faster and further on less fuel. It is also a good conductor of electricity and, since 1900, has been used more widely than copper for this purpose.

Aluminium was deemed by Napoleon III of France to be fit only for himself and his most esteemed guests.

Gallium

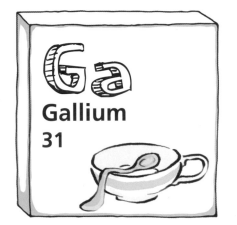

Atomic number	31
Atomic weight	69.723
Abundance	19 mg/kg
Radius	130pm
Melting point	30°C
Boiling point	2229°C
Configuration	(Ar) 3d10 4s2 4p1
Discovered	1875 – P.-E. L. de Boisbaudran

Mendeleev predicted the existence of element 31, which he called eka-aluminium. Just six years later it was found, emitting a distinct violet spectral line, by French chemist Paul-Emile Lecoq de Boisbaudran.

After processing 450kg of ore, de Boisbaudran presented just 600mg of the metal at the French Academy of Science, giving it the name gallium after the Latin 'Gallia' for France. Some say, however, that he hid egocentricity as patriotism, not naming it after France but instead after *'gallus'*, the Latin for rooster, which in French is *'le coq'*. Sadly we will never know if Lecoq's motive was patriotism or immortality.

Water-like

Gallium would melt on a hot summer's day, just below 30°C, forming a very stable liquid. The metal boils at around 2200°C, which gives it the largest temperature range of any known liquid element. Gallium is one of just three elements (the others being bismuth and antimony) whose liquid form is denser than the solid. The solid metal will float on top of liquid gallium. Although we are used to seeing water do this, it is a very rare thing. Today it is known that gallium surrounds itself with more of its neighbours in liquid state than as a solid, but the exact reason for this still remains obscure.

Semiconductor materials containing gallium arsenide harness the power of the sun in solar cells and release light from light-emitting diodes (LEDs).

Solar cells and LEDs

Gallium metal is not really used for much, but as the crystalline semiconducting compound gallium arsenide it finds its greatest use. When an electric current is passed through it, it emits a range of different light, and is used to make light-emitting diodes (LEDs). It is also more efficient than similar silicon technology at harnessing the energy of the sun in solar cells used on solar-powered vehicles and satellites.

Indium

In
Indium
49

Atomic number	49
Atomic weight	114.818
Abundance	0.25 mg/kg
Radius	155pm
Melting point	157°C
Boiling point	2072°C
Configuration	(Kr) 4d10 5s2 5p1
Discovered	1863 – F. Reich & T. Richter

The name indium originates from the indigo blue spectral line that disclosed its identity to German chemists Ferdinand Reich and Hieronymous Richter in 1863.

Soft and sticky

Indium is a soft metal that remains malleable even when cold. It also sticks to itself and other metals easily. These two properties make it excellent at joining two metals when fabricating specialist low-temperature lab equipment. The element's stickiness is also used in solders that bind pieces of metal together with molten metal. Despite being soft, indium adds strength to other metals when alloyed: adding small amounts of the metal makes gold and platinum much harder. Indium alloys are used to make aircraft parts where a large amount of wear is expected.

Seeing clearly

In 1924 there was a little over a gram of pure indium globally. Today over 600 tonnes of indium are refined each year and an equally large amount recycled. Around 45% is used to make indium tin oxide (ITO), a compound which is transparent to visible light and able to conduct electricity. It takes electric current to liquid crystals or LEDs to create the pictures on all modern-day displays.

ITO powers everything from smartphone screens to TVs, and its demand rises with our insatiable appetite for technology. When painted onto architectural glass in warm countries it keeps buildings cool: as it lets the visible light in, it blocks the warming infrared light. ITO is also used on aeroplane and car windows that can be electrically heated to melt frost and ice.

Indium's importance has seen indium prices increase dramatically. Improved recycling extraction efficiency is currently maintaining a good balance between demand and supply, but some countries are stockpiling the metal to make electronics.

Transparent but electrically conducting indium tin oxide has been central to the touchscreen revolution.

Thallium

Tl
Thallium
81

Atomic number	81
Atomic weight	204.389
Abundance	0.85 mg/kg
Radius	190pm
Melting point	304°C
Boiling point	1473°C
Configuration	(Xe) 4f14 5d10 6s2 6p1
Discovered	1861 – W. Crookes

The 'poisoners poison', this element is historically infamous as a murder weapon. William Crookes, at the Royal College of Science in London, was first to notice this element emitting a distinct green spectral line from impure sulfuric acid.

Making a murderer

In her 1961 novel *The Pale Horse*, Agatha Christie wrote of a character attempting murder that his weapon of choice was a thallium poison. Christie's accurate description of thallium poisoning in the book is thought to have saved lives, and contributed to the arrest and conviction of a British murderer. He had been slipping colourless, odourless and tasteless thallium salts into family and co-workers' tea and managed to kill his stepmother and two colleagues, as well as making 70 more people ill. On a lighter note, after reading Christie's book, a nurse was able to recognise thallium poisoning in a child when doctors were stumped, saving her life.

The effects

Thallium has biological properties similar to potassium and, if ingested, is pumped into cells in its place. Here it wreaks havoc with the essential processes which are otherwise the domain of potassium (see Potassium and Sodium). It is usually

A British man killed three and made many more sick when he added thallium compounds to their tea.

treated with Prussian Blue, with the catchier chemical title of Iron(II,III) hexacyanoferrate(II,III). This passes through the digestive system, absorbs the metal, and passes out the other end.

Still useful

Radioactive isotopes of thallium can be used as medical tracers, capitalising on our bodies' ready absorption of the metal. Thallium-201 is used to image blood flow to the heart in patients suspected of coronary diseases. Thallium sulphide or bromide are used in photocells and sensors because its electrical conductivity changes with the intensity of infrared light.

Tin

Atomic number	50
Atomic weight	118.71
Abundance	2.3 mg/kg
Radius	145pm
Melting point	232°C
Boiling point	2602°C
Configuration	(Kr) 4d10 5s2 5p2
Discovered	3500 BCE

Adding a little tin to copper makes the alloy brass which proves hardier, but easier to mould and sharpen. Any ancient fighter with a brass weapon had a distinct advantage over anyone wielding softer and blunter copper.

Ancient invasions

Because of this potential, tin became an important commodity of the ancient world, to the extent that its trade was a closely guarded secret. Ancient Greeks spoke of 'tin islands', believed to be off the north-west coast of Europe, which mostly likely never existed. All they knew was that tin would arrive from the north-west, probably from mines in Spain or England.

The English tin mines of Devon and Cornwall were considered to be the reason the Roman Empire ventured into the barbaric British Isles. The Romans called tin *stannum*, and the mines in the south-west UK 'stannaries'. It is this Latin name that gives us the seemingly illogical symbol for the element of Sn.

Tin may have lost Napoleon the 1812 Russian campaign after the tin buttons on his soldiers' uniforms disintegrated in freezing winter temperatures, killing many through hypothermia.

Ringing true

Tin is known to switch between forms (allotropes) at low temperature, from metallic white (alpha) tin to the non-metallic grey (beta) tin. This occurs for pure tin at around 13.2°C, but any impurities would lower this temperature.

Many an organ pipe or bell has fallen prey to 'tin leprosy': they are traditionally made from tin as it is the most tonally resonant of all metals. Those ringing out today are usually a 50:50 mix of tin and lead to avert their demise in winter. It is also thought by some that Napoleon Bonaparte may have lost his 1812 Russian campaign thanks to this allotrope change of tin (see image).

Tin holds the record for having the most stable isotopes of any element, 10 in total (see table). It is generally unreactive which is why cans have their insides coated with the stuff to prevent food reacting with the steel they are made from, lending its name in the process. The metal was also used for a number of other applications but has been surpassed in price or properties by other materials, leaving just its name behind.

Bismuth

Atomic number	83
Atomic weight	208.98
Abundance	0.009 mg/kg
Radius	156pm
Melting point	271°C
Boiling point	1564°C
Configuration	(Xe) 4f14 5d10 6s2 6p3
Discovered	1753 – C.F. Geoffroy

Bismuth has been known of since ancient times: it was alloyed with tin by the Incas to create bismuth bronze to make knives. It was not until 1753, however, that French chemist Claude François Geoffroy showed it to be an element.

Sort of stable

An article in the Journal *Nature* in 1949 started a long debate regarding the stability of the seemingly stable ^{209}Bi isotope. If stable, then ^{209}Bi would be the heaviest stable isotope of any element. But the article, and those that followed, suggested that it is meta-stable; that there is a very small but still finite probability that it would decay, emitting an alpha particle. It was not until 2003 that French astrophysicists were able to observe the isotope's decay. They measured its half life to be over a billion times greater than the age of the Universe at 1.9×10^{19} years. So as far as science is concerned it is radioactive, but for all practical applications we can treat ^{209}Bi as if it is stable.

Not as nasty as the neighbours

Considering the extremely poisonous elements surrounding it on all sides, bismuth compounds are surprisingly safe, many even less toxic than everyday table salt. This has led to their use in many industries: bismuth oxychloride gives a metallic sheen to cosmetics; bismuth nitrate oxide is an antiseptic used in operating theatres; bismuth subsalicylate is used as an anti-diarrhoeal and anti-inflammatory in medicines such as Pepto-Bismol. The field of bismuth research is thriving, with many looking into using the metal for various catalytic reasons, the most successful to date being as a catalyst in organic synthesis.

Low melting, many uses

With low toxicity but physical similarities to lead, bismuth metal is used in many ways as a lead replacement. Lead shot for game hunting is outlawed in many countries and has been replaced instead with non-toxic bismuth metal. When mixed with other metals, bismuth's low melting point is used to fine tune the melting points of alloys, used for solders, safety valves that melt at certain temperatures and the hot metal type once used in printing presses.

Lead

Pb

Lead

82

Atomic number	82
Atomic weight	207.2
Abundance	14 mg/kg
Radius	180pm
Melting point	327°C
Boiling point	1749°C
Configuration	(Xe) 4f14 5d10 6s2 6p2
Discovered	7000 BCE

Another one of those elements with an illogical symbol, lead owes Pb to its Roman name, plumbum. The same name gives us the English words for plumbing and plumbers as it was used extensively by the Romans for domestic water pipes. They used it because it was a soft, easily shaped material, despite their knowledge of the effects of lead poisoning.

Madness and mayhem

Lead has no known function in the body. If it gets in then it just gets in the way of a wide range of important processes. It is usually found lodging for weeks in the blood, months in soft tissue and years within teeth and bone. It can take out enzymes such as ALAD (delta-aminolevulinic acid dehydratase), essential in synthesising the key part of haemoglobin (see Iron). The brain is most sensitive to lead as it is mistaken for calcium and pumped directly into neurones. It distorts these nerve cells and prevents them from firing and receiving signals from others. This leads to all manner of cognitive diseases.

Lead poisoning may have been the downfall of the Roman Empire. The sugary compound lead(II) acetate was added to sweeten wine, leading to madness among the upper classes. It is still a hotbed of debate as to whether this contributed to the downfall of the Empire.

Safe and unsafe uses

Lead salts find safe use for colouring pottery and plastics, so tightly bound among other atoms it cannot escape. As a metal it is used in laboratories to contain strong acids as it is relatively unreactive for a metal.

Probably the most devastating use of lead, from the 1920s to the late 1970s, was as the additive Tetraethyly lead (TEL) for petrol, which smoothed engine running. This spread the metal far and wide resulting in the poisoning of an untold number of animals. It was the invention of Thomas Midgley Jr., who amazingly also invented ozone-eating CFCs!

Lead salts were used by Romans to sweeten their wine, which unfortunately lead to madness and possibly the downfall of the Empire.

Metalloids

The metalloids are neither metal nor non-metal as they show behaviour associated with both. They straddle the diagonal dividing line which splits the p block in two from boron down to astatine.

Like a non-metal

Like non-metals, the atoms of metalloid elements can usually be found in a number of configurations called allotropes. They also predominantly form covalent rather than metal-like ionic bonds between these atoms, sharing their valence electrons rather than giving them up.

Semiconducting

Metalloids do not have as many delocalised electrons as metals but can still conduct electricity. Physicists distinguish between the energy of electrons that are bound to the atom (valence band) and delocalised electrons able to conduct electricity (conduction band). In metals some valence electrons have enough energy to delocalise and find themselves in the conduction band.

Within metalloids, however, there is an energy gap between the valence and conduction band.

These metalloids can only conduct electricity if the valence electrons gain extra energy from somewhere to jump the gap into the conduction band. General thermal energy from the temperature of the surroundings allows a few electrons to delocalise. As their numbers are much fewer than in a metal, their electrical conductance is poorer, which is why metalloids are often referred to as semiconductors.

If you heat or shine light upon a semiconductor you can give valence electrons extra energy. More can then make the jump into the conduction band, allowing them a higher electrical conductance. Electrical insulators, materials that cannot conduct electricity, have an insurmountable energy gap. They are most likely to start to burn or melt before the energy is anywhere near close enough to promote electrons to the conduction band.

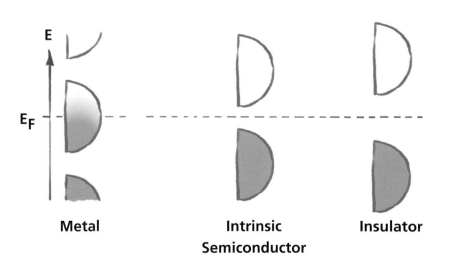

Metal   **Intrinsic Semiconductor**   **Insulator**

Because of their differing desire to hold on to them, various proportions of electrons surrounding metal, metalloid and non-metal atoms are free to move and conduct an electric current.

The secret life of the periodic table

Boron

Boron
5

Atomic number	5
Atomic weight	10.812
Abundance	10 mg/kg
Radius	85pm
Melting point	2076°C
Boiling point	3927°C
Configuration	(He) 2s2 2p1
Discovered	1808 – L. Gay-Lussac & L.J. Thénard

While many elements are named after places, the town of Boron in California is named after the element. The town sprung up around a borax mine, a mineral which lends its name to the element.

Boron is the lightest of the metalloids. Undecided on what it wants to be, alone it forms a number of weird and wonderful formations (allotropes). While the majority of boron is brown and amorphous (without shape), some also exists naturally in crystalline states. The variety of crystalline forms are most obvious from their colour, ranging from clear red to shiny silver-grey and opaque black.

Rocket fuel

But it is when combined with others that boron is really confused. When combined with nitrogen it can form soft powder-like compounds or crystals that rival diamond for hardness. Combined with hydrogen it forms pentaborane (B_5H_9), investigated as a rocket and jet fuel during the Cold War. Although it releases more heat than carbon fuels, research stopped because of its toxicity and tendency to spontaneously burst into characteristic green flames.

Keeping clean

With three valence electrons, each atom is built to make three bonds, e.g. boron trifluoride (BF_3). In this state it can make a fourth bond, taking on a negative charge as it does, before readily breaking a bond to become neutral once more. This crisis of personality means boron is brilliant at passing electrons between atoms as a catalyst. It can also lead to some boron compounds being unstable. Sodium perborate breaks up in warm water, releasing hydrogen peroxide which is used as a bleaching agent in laundry detergent and teeth-whitening products.

Persil takes its name from the sodium perborate and silicate ingredients that get the whites 'whiter than white'.

Silicon

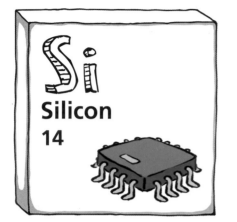

Si
Silicon
14

Atomic number	14
Atomic weight	28.0854
Abundance	282000 mg/kg
Radius	110pm
Melting point	1414°C
Boiling point	3265°C
Configuration	(Ne) 3s2 3p2
Discovered	1825 – H.C. Oersted

The second most abundant element in the Earth's crust, silicon is only beaten to the top spot by oxygen, to which it is invariably bound.

Locked in rocks

Silicate rocks, which give the element its name, occur in an astounding number of varieties. The only differences between them are the metals which find themselves entwined with silicon and oxygen. Most form crystals with repeating units of a silicon atom surrounded by oxygen atoms, connected together directly or with metal atoms spacing them out. They can form long chains which produce brilliant clear or coloured crystals, including everyday glass. Silicate shards are also found in the plant world, for example in the spines of nettles, where they score skin before injecting a mild irritant.

The transistors of computer chips are etched onto single crystal wafers of semiconducting silicon.

Like drawing electricity from a stone

With no metal present, silicon forms the compound silicon dioxide (SiO_2), found in nature as quartz. This and similar crystals emit a small jolt of electricity if squeezed or shocked. This behaviour, known as piezoelectricity, is the heartbeat of digital clocks. In spite of also shining like a metal, silicon is a very poor conductor of electricity, but is very good at trapping electrons.

Computing

Silicon is used as the basis of all modern-day computer processors. By adding small amounts of elements in groups 13 or 15, either side, silicon can be made to conduct electricity in different ways. Processors use these 'doped' semiconductors to trap and release electrons in small box-like structures, which are cut as a lattice into a single crystal of silicon. Each box represents a bit of computer data which can be changed many times each second as it computes some process. It is this that lies at the heart of all modern electronics and has powered the technological revolution.

Germanium

Ge
Germanium
32

Atomic number	32
Atomic weight	72.63
Abundance	1.5 mg/kg
Radius	125pm
Melting point	938°C
Boiling point	2833°C
Configuration	(Ar) 3d10 4s2 4p2
Discovered	1886 – A. Winkler

Mendeleev's most accurate predictions in 1869 were of the properties of eka-silicon, element 32.

Let there be light

Like many metals, germanium is shiny and hard with some uses as a catalyst, yet it is transparent to infrared light and a semiconductor of electricity. Combined with silica it allows infrared signals to pass unhindered through optical fibres which connect the world. Germanium is brilliant at reflecting higher energy light and is used for mirrors in various X-ray applications.

As a semiconductor it rivals silicon for its uses but is not as common, being used mainly in production of light-emitting diodes (LEDs).

Fibre optic cables are produced as hollow glass tubes are filled with germanium tetrachloride gas before being heated to form pure germanium oxide glass in the centre.

Germanium is doped with elements in the adjacent groups to create positive, p-type, and negative, n-type, semiconductors. When sandwiched together a large energy gap is formed in between. When an electric current is applied, electrons gain enough energy to leap across the gap and emit light as they do so.

And dark....matter

Germanium semiconductor detectors are also extremely sensitive to charge atoms or particles. They have found an increased use in airport security around the world to scan for small amounts of radiation in baggage. Germanium detectors are also used in the search for dark matter, which makes up around 27% of our universe but is still to be identified.

The compound germanium dioxide can be used to catalyse the polymerisation reaction producing PET plastic used for drinks bottles, although other procedures are used in Europe and the US. Germanium is named after the homeland of German chemist Clemens Winkler who discovered the element in 1886.

Arsenic
Deadly deceiver

As	
Arsenic	
33	

Atomic number	33
Atomic weight	74.9216
Abundance	1.8 mg/kg
Radius	115pm
Melting point	Sublimes at 616°C
Boiling point	Sublimes at 616°C
Configuration	(Ar) 3d10 4s2 4p3
Discovered	2500 BCE

Arsenic was discovered through a chemical reaction in the 13th century, long before the scientific revolution. Its extraction as an element is attributed to the Catholic bishop Albertus Magnus, who dabbled in alchemy.

Magnus took white arsenic (arsenic trioxide, As_{2O3}), a product from copper refining, heated it in olive oil, and the dark-coloured metallic grey element precipitated out. It owes it name, though, to the Persian word (al) zarniqa which means yellow, a name given to a mineral of arsenic trisulphide used by the ancients as a dye.

Wallpaper killer
Up until Victorian times arsenite salts, with an anarsenic-oxygen negative ion, provided the dyes Paris green and Scheele's green which were unrivalled in their vibrancy. Used extensively in the colouring of wallpaper, and even sweets, it could have been responsible for killing many people in their own homes.

If damp sets in, a particular type of mould (Scopulariopsis brevicaulis) can turn the arsenite into the volatile compound trimethylarsene. This

Arsenic was released by mould growing on 19th-century wallpaper coloured green by arsenite salts. There is a theory that this may have killed the already frail Napoleon Bonaparte in 1821.

molecule is readily absorbed into the body when breathed in. It is debated whether it is this that killed Napoleon while he was exiled in his green-wallpapered room in Saint Helena, in 1821.

Arsenic binds strongly to the keratin in hair; when Napoleon's was tested in 2008 it was found to contain 100 times the modern average arsenic level. As arsenic accumulates over a number of years, this could have arisen from the common 19th century exposure to many other arsenic-containing dyes and glues. Other studies, however, have shown that this arsenic was not in this organic form but instead the mineral form, which suggests deliberate poisoning. Debate rages on.

Biological role

Arsenic binds tightly to the active parts of a whole host of important enzymes, rendering them useless. Most effectively, it disrupts the production of ATP, the molecule which delivers energy to the cells (see Phosphorus). Here arsenate ions compete with phosphate ions. Having cells without energy to perform tasks or repair results in death from multiple organ failure throughout the entire body.

In the 18th, 19th and 20th centuries many arsenic compounds were used in medicines. In the 1910s, German Nobel Prize-winner Paul Ehrlich developed the arsphenamine 'magic bullet' known as Salvarsan to combat the sexually transmitted infection syphilis. This has since been superseded by antibiotics. Over the past 500 years, arsenic trioxide (As_2O_3) has been used to treat cancers and some skin conditions such as psoriasis. At the turn of this century the US government approved the compound to tackle a cancer of white blood cells, acute promyelocytic leukaemia, after it resisted other treatments.

Although used widely throughout history, even in chicken feed as a preventative medicine, use of this toxic metalloid is on the decline.

US soldiers lining up to be injected with the 'magic bullet' arsenic-containing drug Salvarsan to treat the sexually transmitted infection syphilis.

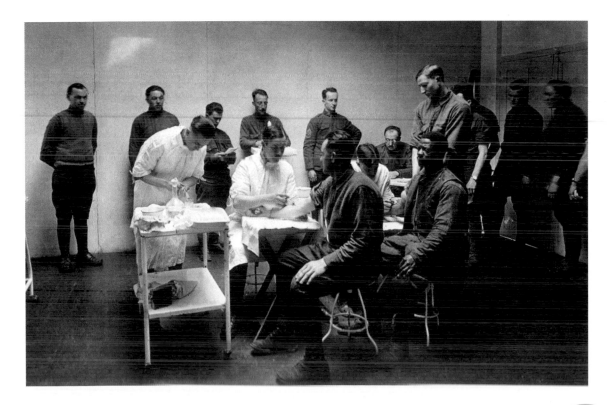

Antimony

Sb
Antimony
51

Atomic number	51
Atomic weight	121.76
Abundance	0.2 mg/kg
Radius	145pm
Melting point	631°C
Boiling point	1587°C
Configuration	(Kr) 4d10 5s2 5p3
Discovered	3000 BCE

Despite being poisonous like arsenic above it, antimony has had a long and fiercely debated history in medicine. It has been known about since around 1600BC, with antimony sulphide used by ancient Egyptians as mascara, and is named from the Greek *anti-monos*, meaning not alone.

An explosive side

As an element, antimony can be found in four different naturally occurring forms (allotropes). Metastable black and yellow allotropes wish to become the shiny grey metallic antimony which is stable, but a poor conductor of electricity. The fourth allotrope, formed by electrolysis, explodes when touched as it also exothermically rearranges into the grey metallic form.

Vomit-inducing

Known today to be poisonous, producing severe liver damage, antimony was prescribed by medical

Mozart was a regular user of antimony tinctures which induced vomiting, which was believed to be the body cleansing itself of ills.

men from ancient Greece and 17th-century Europe. Antimony tartrate was used as an emetic, a vomit-inducer, a process believed to purge the body of other nasty substances. It is believed that a penchant for antimony remedies may have been the cause of Mozart's death in 1791.

Refusing to burn

The greatest modern use of antimony is as the compound antimony (III) trioxide, Sb_2O_3. When this is added to a material alongside halogens then the two work in synergy to act as a flame retardant. At high temperatures the antimony (III) trioxide reacts with the halogens to form trihalides, e.g. antimony trichloride ($SbCl_3$), or oxyhalides, e.g. antimony oxychloride ($SbOCl$). These compounds trap the free radicals released during combustion and prevent a runaway burning of most carbon-based materials. This method is used to produce children's clothing, toys, upholstery fabrics and many other everyday items.

Tellurium

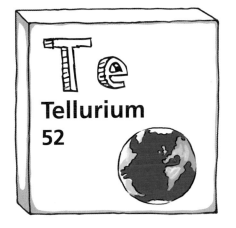

Te

Tellurium

52

Atomic number	52
Atomic weight	127.6
Abundance	0.001 mg/kg
Radius	140pm
Melting point	450°C
Boiling point	988°C
Configuration	(Kr) 4d10 5s2 5p4
Discovered	1783 – F.-J.M. von Reichenstein

Rare tellurium is found today in the slimy residue left over from electrolysis of other metal ores. It was discovered in Sibiu, Romania by Franz-Joseph Müller von Reichenstein from an ore containing gold telluride (AuTe$_2$), in 1783.

Earthly

The article by Reichenstein on the subject, published in an obscure journal, went largely unnoticed until he sent a sample to the German chemist Martin Klaproth in 1796. Klaproth confirmed the discovery and suggested using the name of the only planet in our solar system not represented in the periodic table: tellurium from the Latin *tellus*, or Earth.

Writing its own history

The primary use of tellurium today is as an additive to other metals. When added to steel or copper they become easier to machine, while its addition to lead makes the metal stronger and more durable. As tellurium monoxide, which is actually thought to be a mixture of tellurium metal and tellurium dioxide, it is used in writable optical media such as DVDs. The optical properties of this crystalline compound change notably when manipulated by lasers, allowing data to be written onto a disk. When combined with cadmium, the

Tellurium oxide can be manipulated by lasers to write data to DVD disks.

cadmium telluride semiconductor crystals are more efficient than other similar compounds at converting sunlight into electricity. The future is also looking bright for this element as a main component, in combination with germanium and antimony, in the next generation of phase-changing computer memory chips.

Smelly breath

Tellurium is considered to be mildly toxic to humans, usually involving itself in biological pathways in place of sulfur or selenium. The tiniest dose of tellurium can leave a human with foul breath. Our bodies metabolise the metal in all forms to create dimethyl telluride, $(CH_3)^2Te$, a volatile compound that has a pungent garlic-like smell.

Polonium

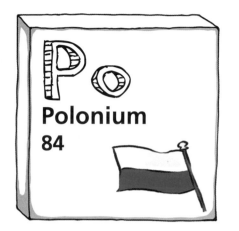

Po
Polonium
84

Atomic number	84
Atomic weight	(209)
Abundance	2x10-10 mg/kg
Radius	190pm
Melting point	254°C
Boiling point	962°C
Configuration	(Xe) 4f14 5d10 6s2 6p4
Discovered	1898 – P. & M. Curie

The 84th slot in the periodic table is a naturally occurring radioelement of 47 isotopes, ranging from 187 to 227. ^{210}Po is the most common, produced in the decay chain of uranium-238. Controversy surrounded the discovery of the element that was tentatively claimed by Marie and Pierre Curie in July of 1898.

For Poland

They wrote 'we believe that the substance we recovered … contains a heretofore unknown metal, similar to bismuth in its analytical properties. If the existence of this new metal is confirmed, we propose that it be named polonium in honour of the native land of one of us.' (Marie was born and raised in Poland, although she performed most of her scientific work in Paris.)

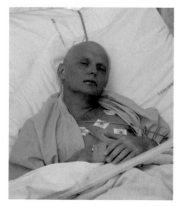

Alexander Litvinenko died in a London hospital on 23rd November 2006, after the only recorded case of polonium poisoning.

The paper sparked the first in a string of debates between the Curies and various German radiochemists. It was argued that the results had arisen from an excited state of bismuth and nothing else. It was not until 1910, after the death of Pierre, that Marie and colleague André-Louis Debierne unequivocally identified the element through spectroscopy.

Death

Polonium's chemistry is little known because of the radioactivity but also the very high toxicity that we associate with other metalloids. This combination makes it one of the most lethal substances known; less than a microgram can cause death. In fact only around 50 nanograms (billionths of a gram) are needed, if ingested, to deliver the mean lethal dose for radiation exposure. It has notoriously been linked to the death of Alexander Litvinenko in London in November 2006. Polonium is also used as the trigger that ignites some designs of atomic bomb.

The secret life of the periodic table

Non-metals
The majority of life, the universe, and everything

Although the periodic table is dominated by five times the variety of metals, there are far more atoms of non-metals in the universe. Hydrogen and helium, classed as non-metals, make up 99% of the entire universe while over half of the Earth's crust, oceans and atmosphere is made from oxygen. Almost all of life on Earth is composed of non-metals, mainly carbon, with only a few metals sprinkled here and there.

Metallic

A non-metal is an element which does not show metallic qualities. Chemical, electrical, physical and mechanical properties are each linked in some way to the behaviour of an atom's valence electrons. Metals are quite willing to give up their valence electrons to form positive ions in ionic bonds. Valence electrons in metals become delocalised, shared between a regular arrangement of these positive metal ions. These free-moving electrons move under the influence of an electric field, conducting electricity effectively. The sea of electrons also leads to a weaker interatomic bond in many metals resulting in relatively low melting points, and making them soft and ductile (able to be drawn out into thin wires).

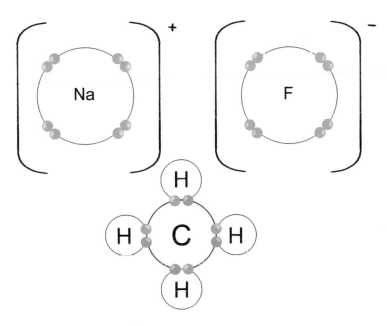

Ionic bonds (top) involve the exchange of electrons between two atoms. They usually form between a metal, that is happy to lose an electron, and a non-metal, which desperately wants to gain one. Covalent bonds (bottom) involve the sharing of electrons and predominantly form between two non-metals as neither one wishes to lose an electron but is more stable sharing electrons with another atom.

Non-metallic

Non-metal noble gases are the most chemically stable group of elements and do not want to react with very much. All elements wish to have the same electron configuration as a noble gas so that they too might find stability. All other non-metals, on the right of the table, are so close to having this that they do not want to let go of the electrons they have and try their best to gain more. They tend to show traits opposite to metals, forming negative instead of positive ions in ionic bonds. They form covalent bonded structures, most of which are small molecules, as they do not merge together through the sharing of delocalised electrons. They are on the whole electrical insulators, hard and brittle, with high melting temperatures.

There are of course exceptions to this generalisation, and the properties within the metal and non-metal classifications vary widely.

Diagonal alley

Keeping a tight grip on their electrons, the lighter non-metals have very high ionisation energies, requiring large amounts of energy to remove the outermost electron. As the size of atom increases, their grip on the electrons weakens as the attractive nucleus is shielded and further from the outer electrons. Electrons become easier to remove, they delocalise into a cloud surrounding positive ions, and the element shows more metallic properties. Because of a decrease in atomic size from across a period, left to right, and the increase down a group, this divides metallic and non-metallic elements by a diagonal line. Non-metals are generally poor electrical conductors and good electrical insulators as their electrons are held tight and do not delocalise.

Categorising further

Non-metals are usually classified into three categories: polyatomic non-metals, diatomic non-metals, and the noble gases (monatomic).

Polyatomic

The polyatomic non-metals of carbon, phosphorus, sulfur and selenium can form two, three or in the case of carbon up to four bonds. With such bond variety they are able to form a wide range of allotropes: different forms of element. With multiple bonds they can form large 3D crystals, 2D sheets or 1D chains. Because of their size these allotropes are usually solid at room temperature.

Diatomic

Nitrogen, oxygen and the halogens of group 17 are all examples of diatomic non-metals. These elements form one or two bonds and are usually found in two-atom molecules with bond(s) tying them together. They do not form larger structures and are therefore mostly found as a gas at room temperature. The exceptions are liquid bromine and solid iodine which experience increased interaction between the larger electron clouds in these molecules.

The noble gases are monatomic, as they do not wish to gain or share any electrons (see Noble Gases).

The diagonal change in properties that arises from the change in ionization energy along a period, and increase in atomic size down a group, results in a stepped line that marks a divide between metallic behaviour and non-metallic behaviour of the elements.

B	C	N	O	F	Ne
Al	Si	P	S	Cl	Ar
Ga	Ge	As	Se	Br	Kr
In	Sn	Sb	Te	I	Xe
Tl	Pb	Bi	Po	At	Rn
Uut	Fl	Uup	Lv	Uus	Uuc

The secret life of the periodic table

Carbon

A chemistry all to itself

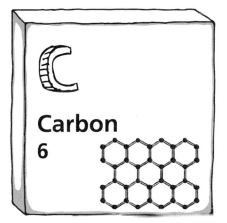

Carbon

6

Atomic number	6
Atomic weight	12.0112
Abundance	200 mg/kg
Radius	70pm
Melting point	3527°C
Boiling point	4027°C
Configuration	(He) 2s2 2p2
Discovered	3750 BCE

Carbon is extremely special, so much so that it has its very own field of science – organic chemistry. The reason for this accolade is the element's ability to form bonds with itself; it is the ultimate polyatomic non-metal, and it has the greatest variety of compounds and allotropes. You can connect tens, hundreds or even thousands of carbon atoms together creating more complex compounds, or an array of elemental forms.

Four carbon bonds – diamond

Ever wondered why steel bridges have that triangular shape? Because triangular structures spread force most evenly. The repeating tetrahedral structure of a diamond is nature's ultimate version of this design. Any force applied to one atom becomes equally shared between the four other carbon atoms it's bonded to, which then in turn distribute their burden to four more.

Natural diamonds are formed deep in the earth as creating this allotrope requires huge temperatures and pressures. Those found have been thrust to the surface in geological upheavals such as a shift in the tectonic plates or volcanic eruptions. In recent years scientists have found ways, using metal catalysts, to artificially

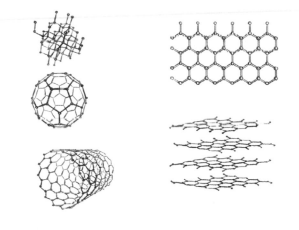

Carbon is able to form the largest number of allotropes: different structures containing atoms of a single element.

manufacture diamonds efficiently in a lab. Man-made diamond can be deposited in thin layers so is used to edge masonry drilling and cutting tools and there is intense research into other uses for this toughest of materials.

Because all four of diomond's valence electrons are bound tightly in carbon-carbon covalent bonds there is no room for them to move. This means electrons cannot absorb any light and jump to higher levels, making it totally transparent. There are also no electrons that can participate in the movement of electric current, making diamond the best electrical insulator known. Thermal vibrations are passed through the lattice effectively which means it is also excellent at conducting heat.

Three carbon bonds

Allotropes containing carbon atoms bonding with three others form repeating hexagonal patterns. Because there are only three bonds being made the fourth valence electron of each carbon atom is shared in a delocalised cloud. This allows these allotropes the ability to conduct electricity, to a varying degree. Delocalisation of electrons also allows these forms of carbon to absorb light.

A collection of irregularly sized random sheets just one atom thick makes graphite – which flakes off onto paper from the centre of a pencil. It looks shiny black because the delocalised electrons are free to go between a vast array of energy levels, absorbing and emitting light of almost any wavelength. Take some sticky tape and peel off just one of these sheets and you have graphene, touted as the 21st century's wonder material. Just one atom thick, the delocalised electrons above and below promise opportunities of creating carbon-based electronics and computing, though as yet there are no practical applications for this form of carbon. Its discovery in 2004 as a new allotrope of carbon won Andre Geim and Konstantin Novoselov the 2010 Nobel Prize in Physics.

The same way that one can make a 3D model from a 2D sheet of paper, these 2D carbon sheets can be formed into a vast array of 3D allotropes.

When folded entirely in on themselves they can form spheres, much like a patchwork football. The shapes of each form different restrictions upon the movement of electrons and so they absorb light of different wavelengths. C70 looks reddish-brown in natural light while C60 is a fantastic magenta.

Rolling the graphene sheets up forms carbon nanotubes. These tiny straw-like structures can be made open at both ends, or with one or both ends closed. Delocalised electrons are able to move along these long tubes very effectively and it turns the relatively poor conductance of graphite into something which approaches a metal.

One or two carbon bonds

Carbon atoms provide the scaffolding for almost every enzyme and organic molecule. In these molecules most carbons bond to one or two more to form chains of varying length. Occasionally they will also form hexagonal rings, similar to those seen in graphite. Various elements bond to carbon, but hydrogen does so predominantly because of its abundance. Each different form provides a different function, and the study of these molecules is the field of organic chemistry. Because of the vast array, there are a few rules to follow when identifying and naming these compounds.

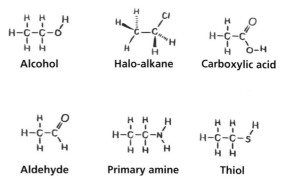

Example functional groups

Alcohol   Halo-alkane   Carboxylic acid

Aldehyde   Primary amine   Thiol

The field of organic chemistry is wide and varied. Here we summarise some of the more common classifications of sub-group which attach themselves to hydrocarbon chains to form all sorts of organic molecules.

Those chains made from just carbon and hydrogen are called hydrocarbons and form the basis of all fossil fuels and plastics. Organic compounds are classified by the functional group subunits that attach themselves to these hydrocarbon chains. Attaching an oxygen and hydrogen (C-O-H) forms an alcohol, such as intoxicating ethanol. Attaching an additional oxygen to the same carbon atom via a double bond (O=C-O-H) gives a carboxylic acid, such as acetic acid – the vinegar you put on your food. Add a nitrogen and you get names like amines and amides, add sulfur and they are called thiols and thials. Addition of nitrogen, sulfur and phosphorus in varying amounts creates various amino acids which are the building blocks of proteins and nucleic acids (DNA).

Further classification arises when considering the structure of the carbon-carbon bonds themselves. A hydrocarbon with all single carbon-carbon (C-C) bonds is an alkane and these have names ending in –ane. If there is a double carbon-carbon bond (C=C) anywhere in a hydrocarbon it is an alkene, with the –ene suffix for naming molecules. If there is a triple carbon-carbon bond then the hydrocarbon is an alkyne lending the suffix –yne. Double-bonded –enes or triple-bonded –ynes can be put through a process called polymerisation. They detach their redundant bonds between the carbon atoms and use them to connect to other carbon-based molecules. Ethene molecules $H_2C=CH_2$, for example, can connect to

each other to form long chains of essentially any length, forming the plastic polythene.

When hexagonal rings form from just carbon atoms, with alternating double single bonds, then you get an aromatic compound. Benzene is the most basic of these, a hydrocarbon, consisting of just a single hexagonal ring attached to hydrogens. The name 'aromatic compound' arises from the odorous early benzene variants, but not all aromatic molecules smell. If you add another polyatomic element, e.g. nitrogen, into the ring you get heterocyclic compounds. Nucleic acids (DNA, RNA), vitamins and steroids are just a few examples of heterocyclic compounds.

Then there are small molecules below 1000g/mol, or with total atomic weight number of each atom adding up to less than 1000. These are also biologically active, like the small and heterocyclic molecule caffeine which may be keeping you awake right now. In enzymes, a number of organic molecules are coordinated by central metal ions. The ions change the shape of the organic molecules to fit other certain molecules and speed up biological reactions (see Transition Metals).

Essential excitement

Carbon is so abundant on Earth because it is a key step in the chain of fusion building heavier atomic nuclei in the heart of a star. But when this field of nucleosynthesis was starting out it came up against a seemingly insurmountable problem with element 6. It did not seem probable that the amount of carbon-12 we see today could have been created through fusion. British cosmologist Fred Hoyle said that in order to create the stable nuclei of 12C that we see in the world around us, there had to be a high-energy, excited state, of carbon. This was an evocation of the anthropic principle, a philosophical idea that suggests that the universe around us exists the way it does because we are here to view it.

The excited state of carbon was subsequently seen in experiments here on Earth.

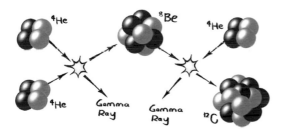

The triple alpha process where three high-energy helium atoms fuse to form an excited, highly energetic, atom of carbon-12.

Phosphorus
Urine luck

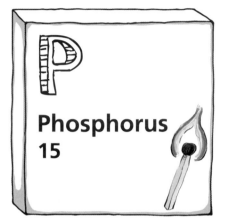

Phosphorus
15

Atomic number	15
Atomic weight	30.9738
Abundance	1050 mg/kg
Radius	100pm
Melting point	44°C
Boiling point	277°C
Configuration	(Ne) 3s2 3p3
Discovered	1669 – H. Brand

German merchant Hennig Brand bankrupted himself, and two wives, searching for the mythical Philosopher's Stone: a material which could turn lead into gold. With no money, he experimented with what he had.

P by symbol, discovery by pee

In 1669 he evaporated the water from his urine and continued to heat the residue until it was red-hot. Condensing the vapour that was given off, under water, he was left with a white powder. On exposure to air this substance burst into bright white flames. Brand had unwittingly become the first person to discover a new chemical element.

He gave the powder, and therefore element, the name 'phosphorus' from the Greek phrase for bringer of light, phosphoros.

Nasty and noxious

This allotrope, known as white phosphorus, has a dark past, having been used in tracer bullets, incendiary bombs, and smoke grenades in 20th-

Incendiary bombs fall toward the city of Dresden, Germany during World War II. Phosphorus is still used today for such bombs as it bursts into flames in the presence of air.

century warfare. In July of 1943, at the height of World War II, 25,000 tonnes of phosphorus bombs were dropped on Hamburg, the city where it was first discovered. They caused widespread death and destruction. More sinister is the use of phosphorus in making chemical weapons such as sarin. This nerve gas interrupts signals sent between nerve cells with devastating effect. It was used by Iraq against Iran in the 1980s and released by terrorists in a Tokyo subway in 1995, killing 12 people and harming nearly a thousand others.

Essential and energetic

Just over 100 years after Brand's discovery it was found that a greater yield of phosphorus came from bone rather than urine. Both sources show just how ubiquitous this element is to life, in the form of phosphate PO_4^{3-}. Phosphates play a key role in the structural framework of many important organic molecules, including one which codes all life on Earth: deoxyribonucleic acid, or DNA for short. Phosphates are found in large quantities within bones which use calcium phosphate salts to stiffen them.

Phosphates are also responsible for delivering energy to every cell in our body in the form of the molecule adenosine triphosphate (ATP). Oxygen is used in respiration to release energy from glucose (see Oxygen). This energy is passed via electrons down a chemical conveyor belt that terminates with the molecule adenosine diphosphate (ADP). Here the energy is enough to attach an additional phosphate group to the adenosine to create the higher-energy molecule ATP. ATP is then transported around the body and delivered to cells. Inside cells it is converted back into ADP, releasing energy that is used by the cell to perform other controlled chemical reactions. The ADP is then recycled back into ATP to be used once more. An average human adult will synthesise their own body weight of ATP each day through the process of respiration. Without this molecule nothing inside our bodies would have the energy required to go on living.

Uses

Along with nitrogen and potassium, phosphorus forms the bulk of all plant fertiliser, the production of which is the primary use of the element in industry. The red phosphorus allotrope is not as flammable as the white and has been used for years on the heads of non-safety matches, as it will burst into flames when struck on almost any surface. Calcium phosphate is used in the production of fine china; it is usually extracted from bone, which explains the term bone china. Tributylphosphate dissolved in kerosene is used to extract uranium from spent nuclear fuel in the Purex process.

A phosphate group added to a molecule of adenosine diphosphate (ADP) stores energy to be released at a later time when the group is removed. This process is utilised to transport energy to cells around our bodies.

Sulfur

Hell's smell and essential acids

Atomic number	16
Atomic weight	32.062
Abundance	350 mg/kg
Radius	100pm
Melting point	115°C
Boiling point	445°C
Configuration	(Ne) 3s2 3p4
Discovered	1669 – H. Brand

Prehistoric sulfur is found in many forms in nature because it forms the most solid allotropes of any element: 30 in total. It is a yellow solid in its most common allotropic form of cyclooctasulfur S8, which melts to a blood-red liquid around 115°C, and burns with a blue flame. Sulfur also forms rings with the number of atoms connected varying from 5–20.

Acid rain

Fossil fuels are the result of dead remains of plants and animals being heated and compressed over millions of years. Sulfur is found within amino acids that combine to make proteins and enzymes found in all life forms. Coal, oil and gas therefore come with extra baggage of sulfur, as a number of other elements. On burning, sulfur dioxide (SO_2) is produced alongside other gases. When this comes into contact with water it reacts to form sulfuric acid (H_2SO_4) or sulfurous acid (H_2SO_3). This occurs within clouds and the resulting acid falls as acid rain. To prevent this from happening, liquid and gas fuels are pre-processed before burning to extract as much sulfur as possible. Solid coal power stations process the gases after combustion to remove sulfur before releasing into the atmosphere.

Sulfur oxides form acid when dissolved in water. With water vapour present in the air they fall as acid rain which destroys stonework and poisons land.

Stink

The Bible mentions that the 'the Lord rained down burning sulfur on Sodom and Gomorrah' and that 'sinners will find their place in a fiery lake of burning sulfur'. Such imagery resonated with the idea that fire brimstone smelt of certain sulfur compounds. The 'smell of hell' comes from the large amounts of hydrogen sulphide (H2S) surrounding active volcanos. Without any oxygen, this is a reduced form of sulfur which interacts quite horribly with our olfactory senses.

Other organic compounds contain the same reduced form of sulfur with the same smelly effects. Thiols (see Carbon) smell so bad that they are added to otherwise odourless methane, propane or butane to act as a stink alarm for gas leaks. Thiol compounds are used by skunks to defend themselves against predators.

Sulfates are found in all sorts of food and drink and some bacteria have evolved to use them as a source of energy. By reducing sulfates to hydrogen sulphide they release energy, much in the same way that we gain energy when reducing oxygen to water (H_2O) in respiration. This leads to stomach-churning smells coming from bad beer, stale wine, and most notably rotten eggs, which most people associate with this form of reduced sulfur.

Uses

Sulfuric acid is the most-used chemical in industry and 85% of the element produced each year finds its way into the stuff. It is a precursor to many other important industrial chemicals that are used across almost every industry. So important is it, that the amount produced is seen as a good indicator of a country's industrial strength and quality of life. It links directly to gross food production with its primary use in the manufacture of fertiliser and is also essential in the processing of wastewater.

Smelly sulfur-containing thiol compounds are used by skunks as a form of protection.

Selenium

Se
Selenium
34

Atomic number	34
Atomic weight	78.971
Abundance	0.05 mg/kg
Radius	115pm
Melting point	180°C
Boiling point	685°C
Configuration	(Ar) 3d10 4s2 4p4
Discovered	1817 – J. Berzelius & G. Gahn

Often overshadowed by its neighbour sulfur, the stench of organic selenium compounds certainly comes out on top.

Electric

The most you might get out of some chemists is that selenium is like sulfur but not as interesting. It has a number of allotropes, the most common of which are coloured black, red or grey. The most prevalent grey form contains large chains up to 1000 atoms long which eventually connect up to form rings.

With a little energy added in the form of light, the electrons within the selenium-selenium bonds become delocalised, and the material is able to conduct electricity. This property was put to use in early photocell light detectors.

Supplement

A lot of people take daily selenium supplements because a number of studies suggest a possible role in cancer prevention, although the evidence is at present inconclusive. Proponents claim it boosts the antioxidant properties of vitamin E, reducing the number of DNA-breaking oxygen free radicals. There is still more research that needs to be done if a definite link is to be found or refuted.

Selenium is required by the body only in

A single brazil nut can provide you with your total daily recommended dose of selenium.

small amounts and is found in a range of foods, particularly nuts, tuna and lobster

Confusion

Both Jöns Jacob Berzelius and Johan Gottlieb Gahn in Sweden made money producing sulfuric acid for industry from ore they mined. In 1817, each noticed that a side product of this manufacturing was a strange red precipitate. When burned it gave off a smell similar to compounds of tellurium. In 1818, Berzelius realised that tellurium compounds were not found in his mine and claimed that the substance must contain a new element. He suggested the name selenium after the moon (the Greek 'selene') because tellurium was named after the Earth (the Greek 'tellus').

Nitrogen

Atomic number	7
Atomic weight	14.0072
Abundance	19 mg/kg
Radius	65pm
Melting point	-210°C
Boiling point	-196°C
Configuration	(He) 2s2 2p3
Discovered	1772 – D. Rutherford

Nitrogen makes up approximately 78% of all the air around us, yet it was discovered more than 100 years after the other elements in its group. It took so long to realise that it was a unique element as it was confused with other gases.

Lifeless

Carbon dioxide (CO_2) had been discovered released from fossil fuels and carbonate rocks. When animals were placed in an atmosphere containing only this gas they perished. Nitrogen, with the same effect, was assumed to be the same thing. This was until the 1760s when Henry Cavendish passed a mixture of this 'mephitic', meaning lifeless, air through an alkaline solution, removing the carbon dioxide, leaving only nitrogen. He correctly noted that the remaining gas had a density just less than air, but he did not publish any of his findings. Instead, credit for the discovery goes to Scottish scientist Daniel Rutherford who performed a similar experiment and wrote of it in his 1772 thesis.

Strong links

Nitrogen is found in air as the diatomic molecule N2 which has the strongest known bond between two atoms of the same element. They share three electrons in a triple covalent bond to create this low-energy stable molecule. Many explosives contain nitrogen atoms isolated in compounds. When the nitrogen atoms come together and form stable N_2 the overall energy is much lower, which leaves a huge amount of energy to be released – the explosion.

Life-giving

Nitrogen plays an essential role in life as a key ingredient in amino acids, which take their name from the amine group -NH_2. These molecules bond together to form proteins which are used in most biological processes in the body. They are also found in nucleic acids, which encode life when chained together to make DNA. We are approximately 3% nitrogen by weight; it is the fourth most abundant element in our bodies after hydrogen, carbon and oxygen.

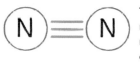

The triple bond binds nitrogen atoms together more securely than any other same element bond.

Oxygen
Double magic

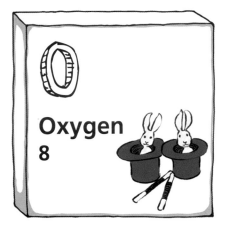

Atomic number	8
Atomic weight	15.9992
Abundance	461000 mg/kg
Radius	60pm
Melting point	-219°C
Boiling point	-183°C
Configuration	(He) 2s2 2p4
Discovered	1774 – J. Priestley & C.W. Scheele

The name of this element comes from the Greek *'oxy genes'* which means acid-forming, which should tell you a little about its reactivity. It is this desire to connect to another atom, and its vast abundance, that makes oxygen essential to our planet.

Oxygen is the most abundant element on the Earth's surface. It is also the third most abundant element in the universe, after hydrogen and helium. It is formed in stellar nucleosynthesis through fusion of nuclei in more stages and across a wider range of star sizes than any other element. This is because oxygen nuclei are doubly magic.

Filling a nucleus

In the introductory chapters we discussed filling electron energy shells and how this relates to an element's chemical reactivity. Noble gases, with full electron shells, are the most chemically stable and inert. When nucleons (protons and neutrons)

Electrons fill energy levels surrounding an atomic nucleus, and neutrons and protons fill energy levels within it. An isotope with full neutron and proton shells is said to be doubly magic and is much more stable than it would otherwise be.

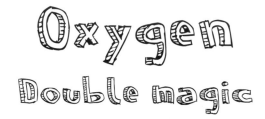

Protons   Neutrons

^{78}Ni

^{48}Ni ^{48}Ca

^{40}Ca

^{16}O

^{4}He

The secret life of the periodic table

are added to a nucleus, such as in nuclear fusion within stars, they too fill energy levels.

Like electrons, these nuclear shells can be defined to first approximation by three quantum numbers. There is an added complication when considering the intrinsic spin of the nucleons. When all factors are taken into account, the first eight nucleon shells can accept the following number of nucleons each: 2, 6, 12, 8, 22, 32, 44, 58.

If a nucleus has either a full proton or full neutron shell then it is magic and more stable than it would otherwise be. Doubly magic nuclei are those in which both types occupy full nuclear shells. Oxygen-16, with 8 protons and 8 neutrons, is the second lightest doubly magic nucleus as both fill the first two nuclear shells completely. The increased nuclear stability of oxygen-16 is the reason for its role in so many stages of nuclear fusion in a such a wide variety of stars.

On Earth

Oxygen is a very reactive element, predominantly due to the two allotropes having unpaired electrons eager to bond so it might achieve noble gas status. On early Earth this locked all of the oxygen up into compounds. Our planet's rocks are about 46% oxygen by weight, much of it in the form of silicon dioxide, which we know most commonly as sand. Because of oxygen's prevalent abundance and reactivity many metals mined from the ground are also found as an oxide compound. Oxygen is also locked up with carbon in the form of carbonates such as limestone. And how could we forget the oceans? They are around 86% oxygen connected to hydrogen as H_2O, water, the best solvent biological life could have hoped for.

In life

Oxygen also combines with carbon to form gases, and the early atmosphere was predominantly carbon dioxide (CO_2). Today oxygen makes up 23% of the air we breathe as the diatomic molecule O_2, and we owe it all to little pioneering bacteria. Through photosynthesis they harnessed the sun's light to fuse water and carbon dioxide together, producing sugars and oxygen gas. Other bacteria used these waste products to release stored energy from the sugars in a process called respiration. Through evolution the ability to photosynthesise was passed onto plants, while both plants and animals respire. Without the waste sugars and oxygen from these bacteria then animals would not have the energy to live and grow.

Two's company

The diatomic form O_2 is interesting because of a couple of unpaired electrons surrounding it. Cool the gas until it becomes a liquid and you will see that it is blue in colour, because these electrons absorb all other energies of light. It is also a magnetic liquid because the electrons align

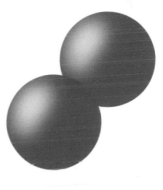

The common diatomic form of oxygen (top) found at sea level and the less stable ozone allotrope (bottom) which is formed through interaction with ultraviolet light in the Earth's upper atmosphere.

themselves against any magnetic field imposed upon them. They also make oxygen very reactive as they will react with almost anything to pair up with other electrons.

Three's a crowd

Ozone, O_3, has an additional oxygen atom and additional unpaired electrons which make it a darker blue as it is able to absorb more light; more magnetic as more electrons align against a field; and more reactive as there are more electrons to pair with. It is created in the cold low pressure of the upper atmosphere from the interaction of O_2 and ultraviolet light. Energetic ultraviolet light splits the bond between oxygen atoms in O_2. The rogue oxygen atoms then combine with another molecule of O_2 to form ozone.

Ozone is essential to life on Earth because is absorbs energetic UV sunlight in the upper atmosphere, preventing it from irradiating the surface of the Earth. This allows organisms to replicate and repair without being overwhelmed by radiation splitting apart molecules like DNA. At ground level, however, ozone is a dangerous pollutant, and is often formed from the burning of fossil fuels. It readily reacts with hydrocarbon molecules, interferes with photosynthesis in plants, and creates toxic smog. The O_3 molecule is higher in energy than O_2 and a single oxygen atom, so it is unstable. If it creeps into the warmer lower atmosphere the thermal energy is enough to break it down, which is why ozone is not usually found naturally in large quantities at ground level.

A time series showing the increase in size of the hole in the ozone layer created by the runaway effects of CFCs.

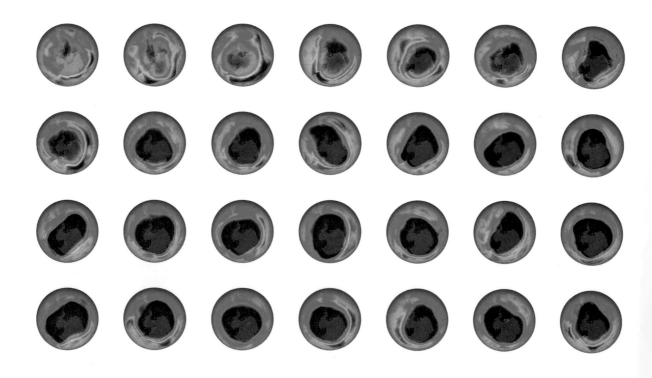

The Halogens

Halogen, translated from the Latin, means 'salt-creating', as these non-metals form ionic bonded salts with most metals. The most familiar salt is sodium chloride, which is the table salt used in food preparation. 'Halogen' had first been proposed in 1811 by German chemist Johann Salomo Christoph Schweigger as a name for the then newly discovered element 17, but it was later named chlorine by British scientist Humphrey Davy.

Green-yellow chlorine gas, red-brown iodine liquid and the lustrous metallic grey of iodine solid, all at room temperature.

Three states of matter

The halogens are the only group in the table whose elements are in all three states of matter at standard temperature and pressure: chlorine, a gas; bromine, a liquid; and iodine, a solid. As elements they all form diatomic molecules, with two atoms bonded covalently. As a gas each halogen is very colourful; from pale yellow fluorine to violet iodine. As solids each halogen has a different look; fluorine can appear as clear or opaque white, while iodine has a dark grey metallic lustre.

Reacting

All group 17 halogens deeply desire just one more electron to be chemically stable, possessing a full valence shell like their neighbours the noble gases. The smaller the atom, the less the positive electrically charged nucleus is screened from attracting an electron. Fluorine, as the smallest halogen, is therefore the most reactive and iodine the least. Astatine is thought to have a similar reactivity to iodine but because it is highly radioactive its chemistry is not very well understood.

Fluorine
The neediest element

Atomic number	9
Atomic weight	18.9984
Abundance	585 mg/kg
Radius	50pm
Melting point	-220°C
Boiling point	-188°C
Configuration	(He) 2s2 2p5
Discovered	1886 – H. Moissan

Fluorine atoms need just one electron to be happy and they are built to get them. Highly reactive as an element, fluorine is used to kill bacteria and sterilise. As the fluoride ion, it is used to provide strength to teeth and treat an array of illnesses in pharmaceuticals.

Very attractive

The further you separate two electric charges, the weaker the force between them; if you double the distance, the force decreases by a quarter. If the two charges are opposite, one positive and one negative, then this force is attractive. A small atom, which holds its electrons tight, allows other atoms to get close to its nucleus. Small atoms, therefore, exert a greater force on nearby electric charges such as electrons in other atoms.

Along a period, from left to right of the table, the size of atom decreases as each additional proton in the nucleus pulls harder on the electrons in the same shell. A fluorine atom is the smallest atom in period 2 which does not have a complete electron shell: just one electron short. This makes it very unwilling to give any of its electrons up, extremely keen to gain an additional electron, and because of its size very able to attract electrons.

Reactive

Fluorine's ability to attract electrons coupled with the desire to fill its shell makes it the most reactive non-metal there is. It reacts with every single element on the periodic table apart from helium and neon, which have an unbreakable hold on their electrons. It is the only element that is able to persuade the heavier and otherwise inert noble gases to react (see Xenon).

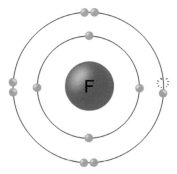

The small fluorine atom is just one electron away from a complete shell and knows how to attract electrons.

Stealing and sharing

Attraction is so great that most of the metal fluoride compounds that form are ionic; fluorine rips an electron clean off a metal atom. Only when metals have already lost a number of electrons, in oxidation states 5+ and over, are they unwilling to lose more and instead share electrons through covalent bonds. Non-metals, as they are close to completing their outer valence shell, are also unwilling to let an electron be taken from them. Most fluorine to non-metal bonds are therefore covalent as well.

Sanitise and strengthen

As the fluoride ion F- in compounds, fluorine has the full set of electrons it wants and finds a large number of safe uses. The most common use is in toothpaste, which contains sodium fluoride or tin fluoride. Our protective tooth enamel is made from calcium-phosphorus compounds called apatites; mostly an apatite which has an OH, hydroxyl, group attached - hydroxylapatite ($Ca_5(PO_4)^3OH$). Hydrogen ions H+ found in acids happily attack this OH group and break apart the apatite, leaving our teeth open to damage.

The fluoride in toothpaste reacts and replaces this OH hydroxyl group in the apatite to form fluorapatite. The stronger bond that the fluoride forms with the apatite means it is not broken down by acids as easily and therefore provides greater enamel protection for our teeth.

Electronegativity

The fluorine-fluorine bond in molecular fluorine F_2 is quite weak, as each atom is constantly looking away from the other for available electrons. As it is so attractive it also forms some of the strongest bonds known between different elements. It forms bonds with carbon that are much stronger than the expected average of carbon-carbon and fluorine-fluorine. In 1932 the American chemist Linus Pauling determined a way of quantifying this in a property called electronegativity. Fluorine is the most electronegative element in the periodic table, so if it combines with another highly

Fluoride ions, added to toothpaste, can replace hydroxide groups in the apatite molecules which protect our teeth, as fluorapatite is much more resistant to attack from acids.

electronegative element then the bond created is extremely strong.

Danger above

Strong bonds formed by fluorine and chlorine with carbon create very stable compounds. These chlorofluorocarbons (CFCs), invented by American Thomas Midgley Junior, were used for many years as a coolant in refrigeration and a propellant in aerosols, as they did not break down under extreme temperatures or pressures. When released these gases circulate to reach the upper atmosphere.

Here CFCs encounter high-energy ultraviolet light, otherwise absorbed by ozone (O_3), which is energetic enough to break apart the strong chlorine and fluorine carbon bonds. This 'photolysis' ($CCl_3F \rightarrow CCl_2F. + Cl.$) produces free radicals: ions which contain unpaired electrons. Radicals attack the ozone to form oxides and diatomic oxygen gas (O_2), but not before creating more oxygen free radicals (O_2-). This chain reaction process led to severe depletion of the ozone layer and to increased levels of dangerous UV radiation bathing areas of the Earth. Use of CFCs has now been banned worldwide and alternates are used: these alternatives still use fluorine's strength in the form of hydrochlorofluorocarbons (HCFCs) and account for 90% of fluorine use worldwide.

Chlorine

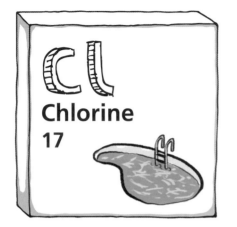

Atomic number	17
Atomic weight	35.452
Abundance	145 mg/kg
Radius	100pm
Melting point	-102°C
Boiling point	-34°C
Configuration	(Ne) 3s2 3p5
Discovered	1774 – W. Scheele

This greenish-yellow gas was first produced in 1774 by Swiss-German chemist Carl Wilhelm Scheele, but he did not realise his achievement, believing it instead to be an oxide of some other element.

In 1810 Humphrey Davy carried out the same experiment, in which hydrochloric acid reacts with manganese (IV) oxide, concluding that Scheele had indeed made elemental chlorine. While the discovery is credited to Scheele it was Davy who named the element, after the Greek for yellowish-green 'chloros'.

Warfare

Chlorine gas is produced industrially in a process developed by German chemist Fritz Haber and was used as the first-ever chemical weapon in World War I. The heavier-than-air gas drifted across the ground and sank to fill trenches. When breathed in, the gas dissolves in water within the lungs to create hydrochloric acid. This corrodes the lungs, stimulating the production of large amounts of fluid which eventually drowns the victim. The armies on both sides had specially trained chemist divisions who were tasked with producing the gas on the battlefield.

Sanitising

The deadly power of chlorine is used to sanitise drinking and swimming pool water worldwide. Its use was first suggested in 1850 by the physician John Snow, after he traced an outbreak of cholera to a single water pump in London's West End. Snow also used the compound chloroform, $CHCl_3$, as an anaesthetic to help Queen Victoria give birth to two of her children. Today chlorine is also used to make bleaches and cleaning products.

Chlorine gas was the first chemical weapon used in warfare, in WWI. Gas masks were issued to soldiers to protect them against its deadly effects.

The secret life of the periodic table

Bromine

Br
Bromine
35

Atomic number	35
Atomic weight	79.9049
Abundance	2.4 mg/kg
Radius	115pm
Melting point	-7°C
Boiling point	59°C
Configuration	(Ar) 3d10 4s2 4p5
Discovered	1826 – A.-J. Balard & C. Löwig

Half a century ago this element could be found in a vast variety of products from fire extinguishers to sedatives. Today it is not as widely used but global production continues to increase as there are a few applications for which bromine cannot be beaten.

Organobromine

The number one use for the element is in brominated flame-retardant products which contain organic bromide molecules. When burnt, these materials produce hydrobromic acid that binds with oxygen atoms, preventing them from finding other fuel to react with. It is also added to plastics to prevent their burning, predominantly in the surrounds of TV screens and laptops which tend to heat up a lot.

Pentabromodiphenyl ether is another flame-retardant molecule which has surprisingly been found in the blubber of whales. These molecules contain the radioactive carbon-14 isotope which confirms that they have come from a biological source. This means that out there in the ocean somewhere there are bacteria making flame-retardants for us.

Smelly

Unlike other elements that have one dominant natural isotope, bromine is a near 50/50 split of

Some Mediterranean molluscs have a purple hue which they owe to a bromine compound – this Tyrian purple was used as a dye for the clothes of Roman Emperors.

^{79}Br and ^{81}Br. It was discovered in 1826 by 24-year-old Antoine-Jérôme Balard after he added acid to sea salt residue from Montpellier, France. An oily red liquid formed that he reported to the French Academy as a new element, which then confirmed his discovery. It takes its name from the Greek word *bromos*, meaning stench, because of its smelly vapours, and is one of only four elements that are liquid at room temperature, along with caesium, mercury and gallium.

Iodine
Essential for development

Atomic number	53
Atomic weight	126.90447
Abundance	0.45 mg/kg
Radius	140pm
Melting point	114°C
Boiling point	184°C
Configuration	(Kr) 4d10 5s2 5p5
Discovered	1811 – B. Courtois

As an element iodine is toxic, but as the iodide ion it is essential for complex life to develop.

Discovery

In the early 19th century, as Napoleon's wars raged on, raw materials used to make the army's gunpowder were in short supply. Where wood was usually burnt for the essential ingredient saltpetre, factories instead began to burn more plentiful seaweed. One such factory was the family business of a young French chemist, Bernard Courtois. Experimenting in 1811, he added some concentrated sulfuric acid to the seaweed ash. To his astonishment Courtois saw a purple vapour released, which then crystallised on the edge of his container. He had discovered a new element which he named iodine, after the Greek for purple, '*iode*'.

Cretinism

Cretinism was an illness that affected many people in central Europe right up to the 19th century. Cretins, those who suffered the illness, showed signs of severely stunted physical and mental growth. Most common in the Alps, cretinism was believed to be caused by stagnant mountain air or bad valley water. English travel guides of the time even remarked on the 'Valleys of the Cretins'. It is now known that this illness is caused by a lack of iodine, or more correctly the iodide ion I- found in its salt compounds, in a person's diet.

Goitre

Iodine levels affect the thyroid gland in the neck, which produces hormones (chemical messengers) that control the growth and operation rate of many systems in the body. A related iodine-deficiency condition, goitre, shows the link with

A lack of iodine results in an under-developed thyroid gland, leading to stunted mental and physical development.

the gland as it causes severe swelling of the thyroid, which is clearly visible in a patient's neck. As iodine is readily taken in by the gland, radioactive isotopes of iodine, such as 131I, can be used as radiotherapy to treat cancer of the thyroid.

Essential for life

The main natural source of iodine is from seawater, with a concentration of 0.05 parts per million (ppm). Those who lived far from the ocean ate plants lacking in, and had no direct access to, this element in their diet. It was just two years after discovery of the element that a pioneering doctor, Jean-Francois Coindet, administered iodine to goitre patients in Geneva who then readily recovered, thus proving this essential link. Today a small amount of iodine salt is added to table salt sold in many countries but related illnesses are less of an issue thanks to a more balanced modern diet. It is not just humans who require iodine. Tadpoles will not grow legs and mature into frogs unless there is iodine present in the water they live in. It is the heaviest element commonly required for life, only beaten overall by tungsten which is used by some niche bacterial enzymes.

Clean

While small amounts of iodide ions are essential for life, the elemental form of iodine is really quite toxic. Iodine tinctures containing just the liquid element alone, or dissolved in ethanol or water, have historically been used as a disinfectant. Cuts and grazes were often painted with the stuff or patients swabbed with iodine before surgery to prevent infection.

When placed in water, iodine dissolves by forming the triiodide ion I3- which is useful for many different chemical analyses. These ions embed themselves within the amylose molecule found in starch where they produce a quite visible dark purple colour. A starch test like this is often done in high school biology to demonstrate areas of a leaf that have produced the molecules through photosynthesis. Counterfeit notes can also be detected if the criminal has printed the fake currency on commercial paper which contains starch.

Tadpoles require iodine in the water in order to mature.

Astatine

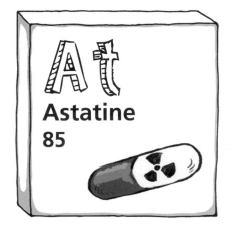

At
Astatine
85

Atomic number	85
Atomic weight	(210)
Abundance	3x10-20 mg/kg
Radius	no data
Melting point	302°C
Boiling point	337°C
Configuration	(Xe) 4f14 5d10 6p5
Discovered	1940 – Corson, MacKenzie & Segrè

Radioactive astatine occurs in nature only as a product of an uncommon decay chain of uranium. It is the rarest natural element known, with less than 50mg estimated to exist in the entire Earth's crust at any one time.

Searching

In 1938 Mussolini's Italy passed anti-Semitic laws banning Jews from holding university positions. Discoverer of technetium Emilio Segrè was visiting Berkeley Lab in California at the time and, being Jewish, he decided to stay. It was here that element 85 became the second element to be discovered from man-made processes, although it was later observed in nature, and the second to be discovered by Sergè. With Dale Corson and Kenneth MacKenzie, he bombarded a sheet of bismuth metal with alpha particles. Bismuth is only two doors down from astatine in the periodic table, so on absorbing an alpha particle it produced astatine-211.

This isotope was found to have a half life (the time it takes for half of a sample of it to radioactively decay) of around 7½ hours. The isotope astatine-210 is the most stable form of the element but has a half life of just 8.1 hours. Its atomic mass and properties neatly fit in the periodic table just beneath iodine. Focussed on the war effort, the Manhattan Project in particular,

Segrè discovered technetium and astatine, but also the subatomic particle the antiproton, the latter earning him the 1959 Nobel Prize in Physics.

the trio did not name the element until 1945, proposing the name astatine from the Greek word '*astatos*', meaning unstable.

Uses

This element has no real use right now but offers promise for future use in medicine to treat and image cancers. The alpha particles emitted when [211]At decays are ideal for targeted radiotherapy to tackle small clusters of cancerous cells. A small secondary decay chain has the added benefit of emitting X-rays which can allow doctors to track exactly where in the body the astatine is.

The Noble Gases

Mendeleev focused on the trends seen in chemical reactions which meant that he was not able to predict unreactive group 18, the noble gases.

Aloof

All the noble gases have a full shell of electrons and therefore have little interest in atoms of other elements, or atoms of their own kind. This results in each of these elements existing as monatomic gases at standard temperature and pressure. Seemingly above the idea of chemical reaction, they are named the 'noble' gases as historically noblemen remained aloof to the common man.

Rare

All apart from argon exist as trace gases in the atmosphere and can only be extracted from liquefied air. Polish scientists Zygmunt Florenty Wróblewski and Karol Olszewski first produced liquid air in 1883. They achieved this through a cycle of compressing the gas, cooling this now hot gas to room temperature, and then cooling it further as it expanded into another vessel. This technique of liquefying air was honed by William Hampson and Carl von Linde who both filed for a patent in 1895. It was Hampson who liquefied the 18 litres of 'argon' that William Ramsay and Morris Travers had collected, and the pair then used to discover the remaining noble gases.

Colourful

All of the gases were identified through their distinctive glow when in an electric discharge chamber. The vivid colours they produced, along with their chemical stability, made them perfect for all manner of electric lighting.

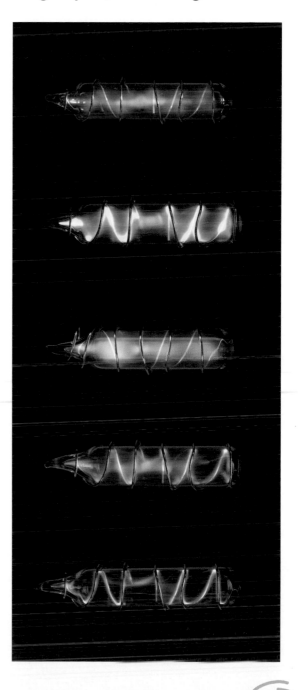

Discharging large electrical currents through noble gases excites them to emit their own characteristic colour of spectral light.
From top to bottom: helium, neon, argon, krypton and xenon.

Argon
First of the nobles

Ar

Argon
18

Atomic number	18
Atomic weight	39.948
Abundance	3.5 mg/kg
Radius	71pm
Melting point	-189°C
Boiling point	-186°C
Configuration	(Ne) 3s2 3p6
Discovered	1894 – Lord Rayleigh & W. Ramsay

It makes most sense to begin with argon as it is the most abundant of the noble gases and was consequently the first to have been found.

Discovery

Shortly after the discovery of argon, there was intrigue around a tiny 0.5% difference in density between nitrogen gas prepared from air and the gas released by ammonia. In 1785 Henry Cavendish wrote of his suspicions that another gas was present in the air. It was not until 1894 that this hunch was proven correct in an experiment carried out by Lord Rayleigh and William Ramsay. They removed oxygen from the air by passing it over hot copper, which readily reacts to form copper oxide. They then removed the nitrogen from the air by passing it over hot magnesium turnings, one of the few metals that reacts with the gas, to form magnesium nitride.

The pure air was then cycled through a spark chamber with oxygen gas to ensure a reaction with any other gases present; nitrogen combines with oxygen, for instance, in the presence of a bolt of lightning. These oxides were then removed from the air by passing it through a weak alkali solution. Excess oxygen was then removed by the hot copper process above. This process was repeated, with minor changes, multiple times until the pair were happy with the small amount of gas they had extracted. Through separation of the gases using their differing rates of diffusion across

Argon is the most common noble gas as it is the final stop in the radioactive decay of heavier elements. The potassium-40 isotope decays to form stable argon-40 11% of the time.

a diaphragm, a process called atomolysis, they unsatisfactorily determined a density of the impure sample that remained. Despite contamination from some nitrogen still present, it was the spectral lines that showed for certain that the gas was a new element.

Common, for a noble gas

As it had not reacted with anything Ramsay and Rayleigh had thrown at it, they named it 'argon', from the Greek *argos*, for lazy. It turns out that this languid gas makes up around 1% of the air around us, and is the most common of the noble gases. Most of this is in the form of the isotope ^{40}Ar (99.6%), with trace amounts of ^{36}Ar (0.34%) and ^{38}Ar (0.06%). The relative abundance is a result of naturally occurring 40K (half-life of 1.25×109 years) found in rock, producing the stable daughter ^{40}Ar in 11% in a process of electron capture. The other 89% of the time it produces radioactive ^{40}Ca via beta decay. Measuring the amount of argon gas trapped within rock allows geophysicists to age rocks in a method known as K-Ar dating.

The predominance of radioactively produced 40Ar is responsible for the atomic weight of argon here on Earth being greater than that of the next element, potassium. This was puzzling when argon was discovered, but later explained by Henry Moseley introducing the atomic number.

Rare in space

Where there is not a collection of radioactive potassium the isotopic content of argon is very different. Argon that is produced in the heart of a star is dominated by the ^{36}Ar isotope. Measurements show that it accounts for 84.6% of all argon ejected from the sun, along with other particles, in the solar wind.

The isotope ^{40}Ar is the least abundant isotope found in the icy outer gas of giant planets of our solar system, which is dominated by the stellar-produced ^{36}Ar which is around 8400 times more common. The atmosphere of rocky Mars contains 1.6% ^{40}Ar, a content similar to Earth. Mercury was shown by the Mariner probe in 1973 to have a very thin atmosphere comprising 70% argon, which is thought to have been released from radioactive decay of rocks on the surface.

Lazy but useful

All of the noble gases find use for their inert nature but the abundance of argon in the air means that it is the cheapest to obtain. Around 700,000 tonnes of the gas is extracted for a range of uses yearly, all from the liquefaction of air. It is used in industry wherever you do not want the product on which you are working to react with the gases in the air. This is mainly at high temperatures, such as in arc welding of aluminium, where every minute 20–30 litres of argon gas surround the large electric current fusing the metal together.

Argon is commonly used to fill the space between panes of glass in double glazing as it is a more effective insulator than air. A little-known use of argon is in the poultry industry. As it is heavier than the main components of air it is a low-lying gas that can fill the lungs of animals, eventually causing asphyxiation. It is used as a more humane method of slaughtering birds, compared to the industry-standard electrified water bath. An added bonus in replacing the oxygen-rich air is an enhancement of the shelf life of the poultry, because bacteria cannot respire and grow.

Unreactive argon gas is used to surround the electrode of a MIG welding torch to prevent the metals involved from reacting with oxygen in the air.

Neon

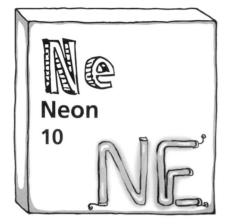

Ne
Neon
10

Atomic number	10
Atomic weight	20.1797
Abundance	0.005 mg/kg
Radius	no data
Melting point	-249°C
Boiling point	-246°C
Configuration	(He) 2s2 2p6
Discovered	1898 – W. Ramsay & M. Travers

Neon was the second noble gas to be discovered by Sir William Ramsay and Morris Travers, in May–June 1898. As liquefied air warmed they siphoned off gases at different temperatures as they boiled, a process called fractional distillation.

Almost argon

One fraction contained a gas which produced a distinctive red glow when excited by electricity. This was an unmistakable hallmark of a new element. The red glow of this neon, from the Greek for new, 'neos', would ignite a lighting revolution. Frenchman Georges Claude's company Air Liquide produced massive quantities of the gas from the turn of the 20th century. In 1910 he tried to sell an idea for neon lighting, where the gas glowed its characteristic red in the presence of electricity. Although nobody wanted to light their house red he soon bent his lighting tubes to form letters and sold the first neon sign to a car dealership in Los Angeles. It did not take long before everyone wanted an eye-catching neon sign to advertise their product.

Important role

Around the same time Englishman J.J. Thompson was experimenting with ions and electromagnetic fields. Using an apparatus called a Crookes tube he stripped electrons from atoms and deflected the resulting ions in magnetic and electric fields. The amount an ion curved in a given field depended upon their atomic weight.

Two patches appeared on his photographic plate, close together, related to an atomic weight close to that of neon. Thomson concluded that some atoms in the neon gas had to have been of higher mass than the rest. Although not fully understood at the time, it was the first evidence that stable isotopes of elements exist: ^{20}Ne and ^{22}Ne.

In the first example of its kind, J.J. Thompson discovered that neon gas has multiple stable isotopes.

Krypton

K r
Krypton
36

Atomic number	36
Atomic weight	83.798
Abundance	1x10-4 mg/kg
Radius	no data
Melting point	-157°C
Boiling point	-153°C
Configuration	(Ar) 3d10 4s2 4p6
Discovered	1898 – W. Ramsay & M. Travers

Krypton is not a planet but a noble gas. It does not react with oxygen, so could not form kryptonite. It has been shown, however, to react with fluorine and hydrogen but only under very high pressures, low temperatures, and after being split apart by ultraviolet light.

Kryptonite cannot exist as no compounds of krypton have ever been made from the inert gas.

Discovery

'Owing to the kindness of Dr W[illiam] Hampson, we have been furnished with 750 cubic centimetres of liquid air, and, on allowing all but 10 cubic centimetres to evaporate away slowly, and collecting the gas… we obtained… 26.2 cubic centimetres of a gas showing the argon spectrum feebly, and, in addition, a spectrum which has, we believe, not been seen before.'

The words of William Ramsay and Morris Travers who went on to '…propose to call it "krypton", or "hidden"' from the Greek. The gas has multiple emission lines of which the pair identified a few as truly unique. It glows with a white-looking light when electricity is passed through the gas in a discharge tube. It is used today to provide high-end lighting for photography and in small amounts in compact fluorescent light bulbs to give a varied spectrum of light.

Detect and measure

Krypton makes up just one part per million of air in the atmosphere. The radioactive 85Kr isotope is a by-product of nuclear fuel reprocessing and is used as a global indicator of clandestine nuclear behaviour. Facilities were detected in the early 2000s in Pakistan and North Korea which, it is thought, were producing weapons-grade plutonium.

Krypton is also used by the International Conference on Weights and Measures who defined the metre as 1,650,763.73 wavelengths of light emitted by the krypton-86 isotope in 1960. This definition changed in October 1983; today a metre is the distance light travels in a vacuum during 1/299,792,458 of a second.

Xenon

A reaction from a stranger

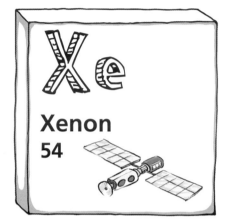

Xe

Xenon

54

Atomic number	54
Atomic weight	131.293
Abundance	3x10-5 mg/kg
Radius	no data
Melting point	-112°C
Boiling point	-108°C
Configuration	(Kr) 4d10 5s2 5p6
Discovered	1898 – W. Ramsay & M. Travers

From their 18 litres of 'argon' extracted from the air, and then liquefied, William Ramsay and Morris Travers also discovered element 54, which they called xenon, from the Greek *'xenos'*, for stranger.

They wrote 'Xenon is very easily separated, for it possesses a much higher boiling-point, and remains behind after the others have evaporated.' The gas was identified by characteristic blue spectral lines, which give the gas its familiar glow in a discharge tube. The pair also stated that, 'Xenon appears to exist only in very minute quantity,' and it does – less than 0.08 parts per million of the air. This rarity makes it a very precious and expensive commodity.

Midday

Xenon is used, like other noble gases, in electric discharge lamps. High-pressure xenon lamps produce a spectrum of light similar to that of the midday sun. They are also very bright. They find use in standard, IMAX and digital movie projectors as well as high-end military lighting.

Feeling sleepy

Xenon gas is an effective general anaesthetic, inhibiting many receptors and channels which pass ions into and out of cells. This effective shutdown was first used on humans in the 1940s but it is

New recycling techniques have made it cheap enough to use expensive xenon gas in hospitals where it can be used to anaesthetise patients before surgery.

The secret life of the periodic table

only now, with improvements in recovering and recycling xenon, that is thought of as economically viable to use.

Sci-future

Heavy and inert, xenon gas is also finding use in ion drives. The spacecraft propulsion systems ionise the gas before accelerating it across an electric field. The heavy atoms mean that a larger momentum can be gained for each atom thrown out of the back of the craft.

Reaction at last

While teaching at the University of British Columbia, Canada, Briton Neil Bartlett changed our idea of noble gases. One of the PhD students he was supervising conducted a reaction in which the gas platinum hexafluoride (PtF6) stripped electrons from oxygen atoms to form an ionic salt. This was astonishing because oxygen holds on to

its electrons tightly, which is evident in its high ionisation energy. Bartlett scanned the periodic table for elements with similar desire to keep their electrons and found xenon.

He then produced an elegant setup in which xenon gas would be released to mix into a separate glass bulb containing the PtF6. This occurred when a magnet was raised and then dropped, acting as a hammer, to smash thin separating glass. Dark red PtF6 gas mixed with the colourless xenon, after some time forming an orange solid. Bartlett had created xenon hexafluroplatinate, the first compound containing a, previously thought to be inert, noble gas. Since this 1962 experiment xenon has been reacted to form an array of other compounds, so too have the noble gases krypton and argon. In 2000, American Chemical Society members voted Bartlett's experiment one of the top ten most important chemistry experiments of the 20th century.

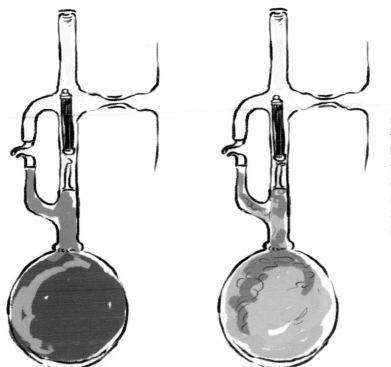

Before and after diagram of the breakthrough experiment by Neil Bartlett, who was able to produce the first noble gas compound, reacting dark red platinum hexafluoride with xenon to produce orange xenon hexafluroplatinate.

Radon

Rn
Radon
86

Atomic number	86
Atomic weight	(222)
Abundance	4x10-13 mg/kg
Radius	no data
Melting point	-71°C
Boiling point	-62°C
Configuration	(Xe) 4f14 5d10 6s2 6p6
Discovered	1900 – F. E. Dorn

Radiation surrounds us, from cosmic rays raining down upon us to heavy elements decaying in the rocks below our feet. We also receive doses of radiation from medical procedures or from industry. None of these come close, though, to the dose of radiation we get from radon gas.

Radon dominates over all of the other sources of background radiation combined. Although drawn in comics as a glowing green, radon is in fact a colourless gas. The gas can accumulate in the basements of buildings in areas with uranium-rich rocks. Detectors can be installed to monitor the danger it presents.

Three become one

In 1900 the German physicist Friedrich Ernst Dorn noticed that radium-containing compounds seemed to release a radioactive gas which he called 'Radium Emanation'. Similar emanations of radioactive gas were seen coming from thorium compounds, in 1899, and actinium compounds, in 1903. These three gases, with abbreviated names of radon, thoron and acton, turned out to be different isotopes of a single element. Naming of elements according to their atomic number forced the IUPAC to name element 86 after its longest-lived isotope, which it turns out is Dorn's radon, which we know today as the isotope ^{222}Rn.

The breaking of DNA due to background radon radioactivity has played an essential part in the evolution of life on Earth.

Evolution

Radioactivity of radon is thought to play a significant role in evolution. Radiation can break bonds in DNA molecules which encode life and after breaking they have a possibility of reforming in a different configuration. While this can lead to cancer, it can also lead to beneficial traits. Those benefitting survived longer and passed it on to a greater proportion of the next generation. Without regular breaking and repair, evolution would have have required much longer to form complex life.

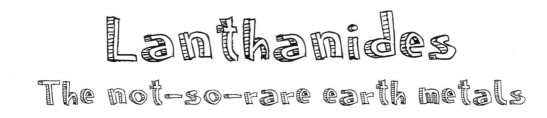

Lanthanides
The not-so-rare earth metals

Get down to period 6 and there seems to be a great divide in atomic number between barium in group 2 and hafnium in group 4. On the common table layout, you usually have to cast your eyes below the main body to find a disembodied block containing two rows of 15 elements each. Seeming like a table all unto themselves, these lanthanides and actinides are actually following the same pattern seen in the rest of the table.

Electron filling

Following the diagonal Madelung rule for filling electron shells, period 5 elements have filled their 5s, followed by their 4d, and finally the 5p subshells. As we go to period 6 the 6s subshell is followed by 4f, a new shape of electron subshell. The f subshells represent higher vibration modes and can each contain 2x(2x3 + 1)=14 electrons (see Quantum Atom for a reminder).

Some forms of the periodic table simply split the table between groups 2 and 3 and place this f-block of elements between. While this form of periodic table is more correct in terms of the quantum structure of the atom, it does require a very wide page. It leads to an extension to 32 groups in total, rather than the traditional table showing 18 with the f-block elements below.

Merging together

The energy of the 4f and 5d orbitals are so similar that in the lanthanides they overlap to the extent that it is difficult to distinguish one from another. This leads to a breakdown of the Aufbau principle, with some choosing to place an electron in the 5d subshell instead of 4f. The new hybrid electron shell also results in some very different chemistry,

s-block

H	He
Li	Be
Na	Mg
K	Ca
Rb	Sr
Cs	Ba
Fr	Ra

d-block

Sc	Ti	V	Cr	Mn	Fe	Co	Ni	Cu	Zn
Y	Zr	Nb	Mo	Tc	Ru	Rh	Pd	Ag	Cd
Lu	Hf	Ta	W	Re	Os	Ir	Pt	Au	Hg
Lr	Rf	Db	Sg	Bh	Hs	Mt	Ds	Rg	

p-block

B	C	N	O	F	Ne
Al	Si	P	S	Cl	Ar
Ga	Ge	As	Se	Br	Kr
In	Sn	Sb	Te	I	Xe
Tl	Pb	Bi	Po	At	Rn

The extended 32-column periodic table places the f-block lanthanides and actinides in their rightful place, showing how the different electron shells are filled.

f-block

La	Ce	Pr	Nd	Pm	Sm	Eu	Gd	Tb	Dy	Ho	Er	Tm	Yb
Ac	Th	Pa	U	Np	Pu	Am	Cm	Bk	Cf	Es	Fm	Md	No

as these elements have more flexibility in where their valence electrons can be to form bonds.

The first row of the f-block is called the lanthanides, after the first element in the period lanthanum. Although lutetium at the end of the row has a full f shell and is starting to fill the 5-d shell, its chemical similarity with the rest of the lanthanides earns it a place with its brothers.

Also called the 'rare earth metals', these elements are not really rare, but certainly behave in chemically similar ways to the group 2 alkaline earth metals. An 'earth' is an obsolete 18th-century term for those elements which form oxides that produce strong alkaline solutions when dissolved in water. Lanthanides also produce basic solutions similar to those of group 2 and share a number of other similar traits.

Not so rare

The abundance of these elements changes by around a factor of 100 as you go from lanthanum to the heaviest in the collection, lutetium. Cerium is the most common and is found as copiously as nickel or copper in the Earth's crust, while lutetium exists in amounts similar to tin. All are over 1000 times more abundant than the platinum group metals in the transition group.

Shrinking

The lanthanides show the best example of the reduction in atomic size as you go along a period, with lutetium about 25% smaller than lanthanum. This reduction in size, known as the lanthanide contraction, is responsible for the third-row transition metals having an atomic size similar to their second-row counterparts.

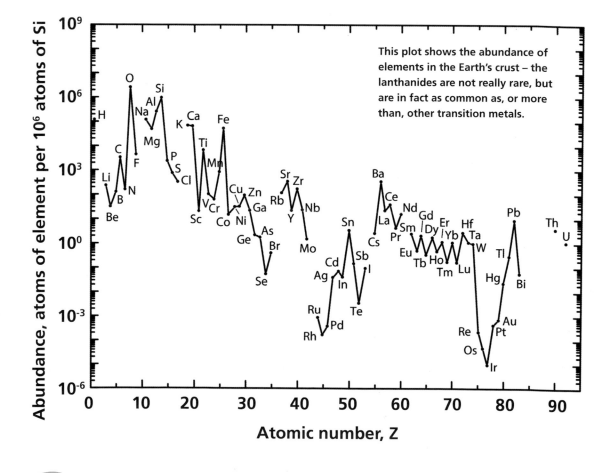

This plot shows the abundance of elements in the Earth's crust – the lanthanides are not really rare, but are in fact as common as, or more than, other transition metals.

The secret life of the periodic table

Lanthanum

La
Lanthanum
57

Atomic number	57
Atomic weight	138.90547
Abundance	39 mg/kg
Radius	195pm
Melting point	920°C
Boiling point	3464°C
Configuration	(Xe) 5d1 6s2
Discovered	1838 – C.G. Mosander

Carl Gustav Mosander discovered a new element hiding within a sample of cerium salts. This sneaky appearance encouraged Mosander to name the element lanthanum, originally lantanium, from the Greek 'lanthanein' meaning 'to lie hidden'.

After decomposing the thermally unstable cerium nitrate to cerium oxide it appeared that 40% of what remained was in fact an oxide of a new metal. With just a weak acid solution cerium oxide does not dissolve, and a solution of the new oxide remains.

The extra density provided by lanthanum increases the refractive index (the ability to bend light and focus it) of glass when made into lenses.

A little bit everywhere

As it is a relatively cheap material to produce, lanthanum finds itself supporting the role of many other elements without ever taking centre stage. It is added to iron and steel in small quantities to make them less brittle. A small amount added to tungsten welding electrodes improves their durability. It can also be found in the mischmetal (German, mixed-metal) that provides the spark in cigarette lighters.

Shining role

As light travels from air into a material it effectively slows down and changes its direction of travel. The amount by which it slows down or bends is a property known as the material's refractive index and is dependent upon the material's density. A dense material places more electrons in the path of the light for it to interact with, slowing it down. Dense metals, when added to glass to increase the refractive index. A higher refractive index bends light more, allowing it to be focussed more effectively. Lanthanum is used as it improves a glass's ability to focus light, but does not produce aberrations: splitting of light into its various colours, such as is seen in lead crystal. Lanthanum glass is used in camera and telescope lenses.

Cerium

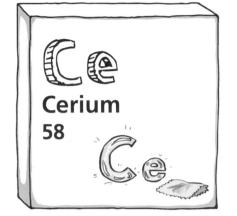

Cerium
58
Ce

Atomic number	58
Atomic weight	140.116
Abundance	66.5 mg/kg
Radius	185pm
Melting point	795°C
Boiling point	3443°C
Configuration	(Xe) 4f1 5d1 6s2
Discovered	1803 – J.J. Berzelius, & W. Hisinger

Cerium is the most common of the lanthanides in the Earth's crust and finds a myriad of uses, mainly in its oxide form, ceria. It was discovered in 1803 by chemistry heavyweight Jöns Jacob Berzelius and Wilhelm Hisinger and named after the recently discovered asteroid Ceres, which takes its name from the Roman God of agriculture.

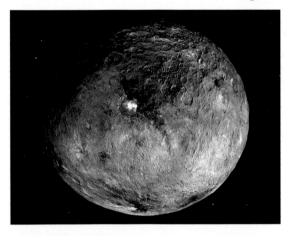

False-colour satellite image of the dwarf planet Ceres, the largest object in the asteroid belt between the orbits of Mars and Jupiter.

Flawed

The main use of this element is in the compound cerium (IV) oxide, commonly known as ceria. While the chemical formula is written as CeO_2 there are pockmarks in the material where oxygen atoms are missing. This is because cerium can be reduced to exist as cerium (III) oxide where each atom connects to an average of 1.5 of oxygen, Ce_{2O3}.

Burning completely

Reduction of cerium (IV) to cerium (III) oxide is used to provide oxygen for a number of different uses. It ensures complete combustion of carbon when added to gasoline and diesel fuel. This releases more energy within the engine and also reduces the amount of toxic carbon monoxide gas emitted from the exhaust. It is also used in self-cleaning ovens where it ensures that any food remains are well and truly burnt from any surface inside a very hot oven.

Colouring and polishing

The cerium (IV) oxide also lends it yellow/gold colour to stained glass, and its hardness is used to grind and polish lenses.

Praseodymium

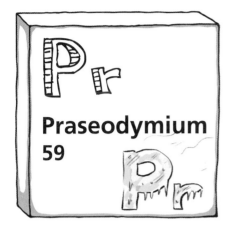

**Praseodymium
59**

Atomic number	59
Atomic weight	140.9077
Abundance	9.2 mg/kg
Radius	185pm
Melting point	935°C
Boiling point	3520°C
Configuration	(Xe) 4f3 6s2
Discovered	1885 – A. von Welsbach

Didymium, from the Greek for twins, *'didymos'*, was the name given to a pair of elements more reluctant than other lanthanides to separate from one another. As a student of Robert Bunsen, Carl Auer von Welsbach untangled praseodymium, green twin, and neodymium, new twin, in 1885 thanks to their slightly different flame spectral lines.

Protecting eyes

In the 1940s it was noted that both of these twins had absorption spectra that effectively filters out the glaring light emitted in glass-blowing or metal-welding. They can both be found today in didymium glass goggles which allow the person working to filter out the unwanted glare and focus on the job in hand.

Sluggish light and cold magnets

Dense praseodymium is also used in a silicate glass which has such a fantastically high refractive index (see Lanthanum) that it slows light down by a factor of a million to 300 metres per second. Praseodymium compounds also make really good magnets, which are used in magnetic chillers. These machines can cool material down to within just a few millionths of absolute zero, around -273°C.

Didymium, a mix of praseodymium and neodymium, absorbs a large amount of the visible light spectrum. It is perfect for use in the glass of protective welding goggles.

Neodymium

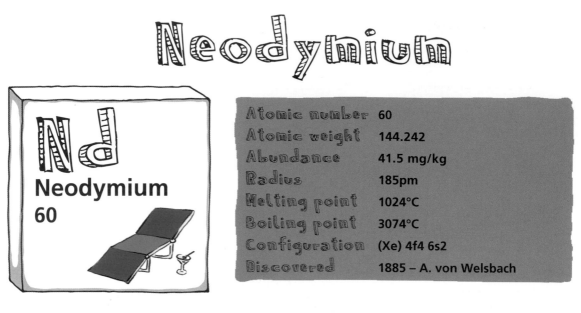

Nd
Neodymium
60

Atomic number	60
Atomic weight	144.242
Abundance	41.5 mg/kg
Radius	185pm
Melting point	1024°C
Boiling point	3074°C
Configuration	(Xe) 4f4 6s2
Discovered	1885 – A. von Welsbach

Different isotopes of neodymium can be found at different depths in the Earth. Looking at the isotopes of neodymium found in lava has been used as an indication of how far the lava has travelled.

Variation in isotopes of neodymium found near volcanoes has enabled geologists to predict the size and scale of possible volcanic eruptions.

Predicting eruptions

Large eruptions have been shown to have very different neodymium isotope fractions than smaller eruptions. Large eruptions show a content commonly seen not in the Earth's crust but from deeper in the mantle. Looking at the neodymium isotopes in lava can help geophysicists predict the size and scale of possible volcanic eruptions.

Large lasers

Neodymium is added in small quantities to crystals of other elements which produces some of the world's most powerful lasers. They produce light in the near infrared and are used by the atomic weapons agencies worldwide to recreate the intense heat and pressure that exists inside a nuclear explosion. Neodymium lasers are also being used to initiate nuclear fusion inside the first test laser-induced fusion reactors, a technology that could lead to inexhaustible amounts of clean energy.

Miniature magnets

An alloy of neodymium, $Nd_2Fe_{14}B$, forms the strongest known permanent magnets. This has enabled the miniaturisation of devices which rely on magnetism. These magnets are used in guitar pickups, earphones and computer hard drives among many other applications. Although cheaper, lighter and stronger than other magnetic alloys, neodymium does not retain magnetism to very high temperatures, limiting its use. Nonetheless neodymium magnet use is on the up with the proliferation of electric hybrid vehicles and permanent magnet wind turbines for clean and renewable electricity.

Promethium

Pm
Promethium
61

Atomic number	61
Atomic weight	(145)
Abundance	2x10-19 mg/kg
Radius	185pm
Melting point	1042°C
Boiling point	3000°C
Configuration	(Xe) 4f5 6s2
Discovered	1945 – J.A Marinsky, L.E. Glendenin & C.D. Coryell

This elusive element is only one of two elements below atomic number 83 that does not have any known stable isotopes, the other being technetium. This led to confusion around its existence and eventual discovery after a number of unsuccessful attempts.

Missing elements again

In 1914 Henry Moseley's concept of atomic number presented a solid prediction of a missing element 61. When ordering the elements according to the number of protons in an atom, it became clear that a gap existed between elements 60 and 62. In 1926, Italian and American groups both claimed to have isolated the element from rare earth minerals.

Statue of the Greek Titan Prometheus at the Rockefeller Center, NYC.

Florentium and illinium

Although the Italian group published their findings shortly after their American counterparts, they claimed to have precedence because they had been sitting on their results for two years. The Italians wanted the name florentium, after Florence where they conducted their work, while the Americans wanted Illinium after their institute, the University of Illinois. Both groups claimed to have seen a unique set of spectral lines but these were later attributed to impurities, mainly didymium.

Fire of the gods

The uranium reactor at Oak Ridge National Laboratory was commissioned for the sole purpose of producing plutonium for the first atomic bombs. Jacob A. Marinsky, Lawrence E. Glendenin and Charles D. Coryell finally discovered element 61 in 1945 when separating material from irradiated uranium fuel to obtain plutonium. After seeing the destruction of the atomic bomb, Coryell's wife suggested naming the element after Prometheus. This Titan of Greek mythology stole fire from the gods and gave it to man, a story she felt was fitting.

Samarium

Sm
Samarium
62

Atomic number	62
Atomic weight	150.36
Abundance	7.05 mg/kg
Radius	185pm
Melting point	1072°C
Boiling point	1794°C
Configuration	(Xe) 4f6 6s2
Discovered	1879 – P.E.L. de Boisbaudran

Samarium's spectral lines were first seen in a sample of didymium in 1853 by Swiss chemist Jean Charles Galissard de Marignac, but not isolated until 1879. With a similar chemistry to other lanthanides, it finds use in geology and technology.

Geologists date rocks

To date rocks, geologists look at the relative amounts of parent and daughter radioactive isotopes. During the rock cycle deposits of uranium and other radioactive isotopes become redistributed, as sedimentary rock turns into metamorphic. This mixing shuffles the parent:daughter ratio, effectively resetting the geological clock. Samarium and neodymium isotopes, however, are highly resistant to redistribution during this metamorphosis process. Because of this they can be used to gauge the age of rocks further back, over time periods greater than the average life of a rock. The comparison of ^{147}Sm to its daughter ^{143}Nd allowed NASA scientists to date rocks brought back from the Moon by Apollo astronauts, and date remains of meteorites here on Earth.

Hot magnet

Samarium compounds form permanent magnets that retain their magnetism to extremely high temperatures. While neodymium magnets are taking over in many day-to-day technologies, samarium-based magnets are used in intense environments such as the magnetrons used in microwave ovens. Samarium magnets also provide superior performance in high-end headphones, microphones and electric guitar pickups.

Samarium magnets are used in the intense environment of a microwave oven magnetron as they retain magnetism to much higher temperatures than most.

Europium

Atomic number	63
Atomic weight	151.964
Abundance	2 mg/kg
Radius	185pm
Melting point	826°C
Boiling point	1529°C
Configuration	(Xe) 4f7 6s2
Discovered	1901 – E.-A. Demarçay

The spectral fingerprint of element 63 was seen twice before a salt of it was produced, and it was definitively discovered by the French chemist Eugene-Anatole Demarçay in 1901.

The British had previously attributed the discovery to William Crookes and the French to Paul-Emile Lecoq de Boisbaudran, but as neither could isolate the element, Demarçay is in the history books. He named the element after, you guessed it, Europe – something no British person of Crookes' time would ever have done.

Seeing invisible light

Europium's presence in the mineral fluorite is responsible for some of its fluorescence, the name given to the glow of this and, now many, other chemicals. Atoms of the element absorb the ultraviolet light invisible to humans and re-emit it as lower energy visible light, usually blue. Europium salts are added to banknotes, including euros, as a security check. When placed under a UV light they will glow blue, wherever deposited, and let the person know that the notes are genuine.

Exciting lighting

Another use for europium is dying from the wave of flat screen LED TVs. Old cathode ray tube (CRT) TVs would fire electrons at a glass screen painted with chemicals that phosphoresce: emit light when excited. Only a phosphor with europium added provided vivid enough red colour for the screens. Europium emits this red light in its plus two oxidation state, but emits blue light in its plus three state. Different salts containing each, along with another greenish phosphor, are used to provide the white light emitted from energy-saving compact fluorescent light bulbs (CFLs).

Different oxidation states of europium produced the red and blue light in old fashioned CRT TVs.

Gadolinium

Atomic number	64
Atomic weight	157.25
Abundance	6.2 mg/kg
Radius	180pm
Melting point	1312°C
Boiling point	3273°C
Configuration	(Xe) 4f7 5d1 6s2
Discovered	1880 – J. C. G. de Marignac

This is the first element we have discussed so far to be named after a scientist: Finnish chemist and geologist John Gadolin, who was famed for extracting the first rare earth elements in the 1790s. It was actually discovered in France, however, almost a century later in 1880, by Jean Charles Galissard de Marignac. Gadolin's family name derives from the hebrew _gadol_ meaning great (see tile).

Magnetic manipulation

Right in the middle of the lanthanides, gadolinium has a huge 7 electrons unpaired in the 4f shell. Easily manipulated, these electrons provide gadolinium with fantastic magnetic properties. It is used as a contrast agent in magnetic resonance imaging (MRI) as it is very clearly seen interacting with the machine's large magnetic fields. Because the Gd^{3+} ion is similar in size to the calcium Ca^{2+} ion it is toxic to many bodily processes if left alone. Instead, the gadolinium is tied up in a complex of other molecules before it is injected into the patient for their scan.

Probing particle physics

Gadolinium is good at capturing neutrons left floating about, and emits a very distinctive wavelength of light when it does. This is being put to use by the Super Kamiokande experiment in Japan as a way to boost the sensitivity of seeing subatomic particles called neutrinos. Some ways in which neutrino particles interact are invisible to traditional technologies. By adding gadolinium salts to the water in the detector, scientists hope to observe these otherwise unseen interactions for the first time.

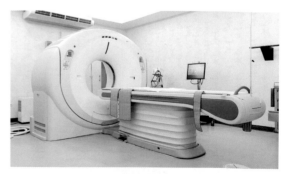

Magnetic Resonance Imaging (MRI) machines image the human body using huge magnetic fields, generated by gadolinium-containing magnets.

Terbium

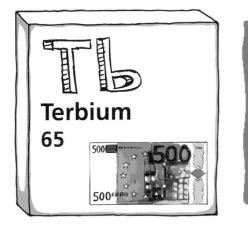

Atomic number	65
Atomic weight	158.9254
Abundance	1.2 mg/kg
Radius	175pm
Melting point	1356°C
Boiling point	3230°C
Configuration	(Xe) 4f9 6s2
Discovered	1842 – C.G. Mosander

Terbium is one of four elements (the others being erbium, ytterbium and yttrium) that were discovered from the feldspar mineral mined in Ytterby, Sweden, from which they all take their name.

Tracing and tracking

In the Tb^{+3} oxidation state found in some salts, terbium emits a distinct green colour when excited. It finds a declining use, along with europium, in the phosphors on CRT TV screens (see Europium).

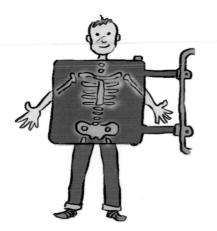

Terbium is also used in active X-ray screens which efficiently convert X-rays into visible light. Here it reduces the exposure time required to get an X-ray image, which reduces the radiation dose that a patient receives.

In biology labs, terbium atoms are tagged onto various molecules and used as tracers, which when excited glow an eye-catching green. This allows the scientist to determine where chemicals end up in biological systems and exactly how they operate. Terbium salts are also added to bank notes where they glow under UV light acting as a anti-counterfeit measure.

Size shift

The terbium-dysprosium-iron alloy Terfonol-D does something really quite strange. Depending upon the magnitude of magnetic field in which it is placed, it changes size, becoming smaller in larger magnetic fields. Used originally in naval sonar systems, it is now found in magnetic sensors, vibrational actuators and sound producing transducers. It has been applied in commercial devices that can turn any hard surface into a speaker. There is a lot of research into how this fantastic magnetostrictive property might lead to microscopic motors or advance fuel injection systems in cars.

Dysprosium

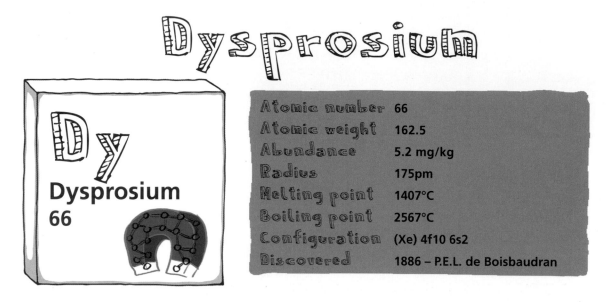

Atomic number	66
Atomic weight	162.5
Abundance	5.2 mg/kg
Radius	175pm
Melting point	1407°C
Boiling point	2567°C
Configuration	(Xe) 4f10 6s2
Discovered	1886 – P.E.L. de Boisbaudran

Dy
Dysprosium
66

Discovery of a number of lanthanides followed a chain of many different precipitations of the oxides of the metal from various solutions. After the discovery, in 1878, that erbium minerals also contained oxides of holmium and thulium, chemists scoured these remnant compounds for any impurities.

Difficult

In 1886 Paul-Émile Lecoq de Boisbaudran's patience paid off. In the fireplace of his Paris home he dissolved a sample of holmium oxide in acid. Carefully adding ammonia in small quantities, he collected a solid which precipitated out. After painstakingly repeating the procedure some 30 times he was eventually left with a sample of a new metal oxide. Because of the effort expended in isolating this new oxide he named the metal bound within it 'dysprosium', from the Greek *dysprositos* ' meaning 'difficult to get'.

Radiation safety

Dysprosium has a number of other applications where you also regularly find other lanthanide elements: in magnetic devices, for neutron capture, and in lighting. It is also mixed into colourless crystals of calcium sulphate or calcium fluoride in safety dosimetry badges worn by radiation workers. The dysprosium atoms become excited by most forms of radiation and emit green light as they come to rest. The light exposes photographic paper, or a digital detector, which is checked regularly to ensure the person wearing the badge has not been exposed to a dangerous amount of radiation.

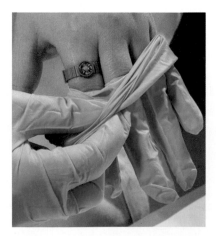

Laboratory dosimeters containing dysprosium track exposure to ionising radiation.

Holmium

Ho
Holmium
67

PHILIPPIUM
MOSANDRUM
HOLMIUM

Atomic number	67
Atomic weight	164.9303
Abundance	1.3 mg/kg
Radius	175pm
Melting point	1461°C
Boiling point	2720°C
Configuration	(Xe) 4f11 6s2
Discovered	1878 – Cleve, Delafontaine & Soret

Holmium is another element with controversy surrounding its discovery. First seen spectroscopically in 1878 by Swiss duo Marc Delafontaine and Louis Soret in Geneva, the discovery was instead credited to the Swede Per Teodor Cleve, who named it after his home town of Stockholm.

Stable spectrum

All lanthanides behave when viewed by a spectroscope. The electron shells of the transition metals, and other elements, distort from their original shape when they form bonds. Lanthanides, it seems, don't really pay any attention to any other atoms they may form a compound with. As their electron shells remain the same shape, lanthanide compounds can be used as a stable reference for anyone carrying out spectroscopy. Holmium was used for many years to calibrate such equipment because it gives a spectrum when in an oxide almost identical to its elemental.

Holmium is used to produce microwave lasers that can cut through flesh.

Microwave cutter

Holmium lasers can produce a wavelength of light close to that of a domestic microwave oven. Such electromagnetic radiation is efficiently absorbed by water molecules because it perfectly excites the hydrogen-oxygen bonds in H_2O. Soft tissue in our bodies is largely made up of water and these lasers are energetic enough to cut through flesh.

Cuts made by holmium lasers are very accurate, within a millimetre's precision. They also have the benefit of self-cauterising, as the heat seals off any blood vessels that it slices through. They are used in a number of different medical and dental procedures.

Erbium
Internet essential

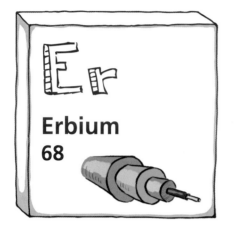

Atomic number	68
Atomic weight	167.259
Abundance	3.5 mg/kg
Radius	175pm
Melting point	1529°C
Boiling point	2868°C
Configuration	(Xe) 4f12 6s2
Discovered	1842 – C.G. Mosander

The gradually changing spectra of the lanthanides provide an array of uses for the rare earths. Erbium is particularly essential to our modern world as it is used in fibre-optic communications.

Optical internet

Signals sent using light can carry much more data every second than those sent down a traditional copper wire. Vast amounts of data are transferred using light, shone down very thin fibres of glass where it remains, for the most part, trapped. It bounces against the sides of the fibre at such slight angles that it rebounds back inside like a skimmed stone upon water, a process known as total internal reflection.

The daytime sky is blue because high-energy blue light scatters more than lower-energy red light.

Why is the sky blue?

Light in the fibres will, though, eventually be lost as it scatters off crystals of silica in the fibres. This Rayleigh Scattering process is also the reason that the sky looks blue in the day and red at sunset. The higher the energy of light, the more it scatters. High-energy blue light scatters more across the entire sunlit daytime sky than any other, which is why it looks blue. When the sun drops in the evening the red light dominates because it takes a more direct route to us, washing out the blue light scattered beyond the horizon.

While some short fibre connections use visible light, not caring about loss from scatter, for longer journeys lower energy near-infrared light is preferred. Although it does not scatter as much, some is inevitably lost after many kilometres. To continue its journey between continents the signal must first be amplified. This is where erbium comes in. Erbium absorbs and emits light of this crucial near-infrared energy.

Light amplification

Sections of erbium-doped fibre are optically stimulated with higher energies of light, putting the atoms in an excited state. When a weak signal arrives, each photon particle of light stimulates the erbium atoms to drop down to a lower energy state. When this happens they emit near-infrared light of the same energy and direction of the original signal. This amplified signal can then continue down an optical fibre for many more kilometres before the whole process happens again.

This technology is responsible for the internet age. Without it we would not be able to consume the vast amounts of data we do, exchanging audio, video and other files worldwide. Thank erbium for the modern internet.

Erbium is also used in infrared photographic filters, used in specialist and mainly astronomical imaging.

Mix-up

Under pressure from his mentor to publish, Carl Mosander published the discovery of terbium and erbium in 1843, despite doubts surrounding the purities of his samples. It turned out that these two samples contained seven new elements in total: erbium, terbium, ytterbium, scandium, thulium, holmium and gadolinium.

In a twist of fate the spectroscopist Marc Delafontaine mixed up samples of the pure erbium and terbium oxides when confirming their existence. This name change has stuck to this day and so Mosander's erbium is terbium and vice versa.

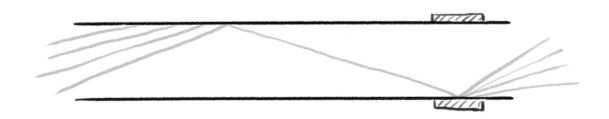

Erbium atoms amplify the signal passing along an optical fibre.

Thulium

Atomic number	69
Atomic weight	168.9342
Abundance	0.52 mg/kg
Radius	175pm
Melting point	1545°C
Boiling point	1950°C
Configuration	(Xe) 4f13 6s2
Discovered	1879 – P.T. Cleve

As you reach the last few lanthanides, the lanthanide contraction slows and they all have a rather similar atomic size. Along with a similar chemistry, and a relative rarity to neighbouring elements, it makes it very tough to get a pure sample of thulium.

Determination

Evidence of the element came from a brightening of distinct green spectral lines as thulium oxide was gradually separated from erbium oxide. It took until 1911 and one very determined Englishman, Charles James, before the first pure sample of a thulium compound was extracted. Working from the University of New Hampshire, USA he prepared a bromate from an impure sample of erbium oxide which faintly showed thulium's spectral signature. He dissolved the bromate in alcohol and extracted coloured compounds that crystallised out from the solution. Dissolving these bromate crystals in alcohol once more, he repeated the same process

It took chemist Charles James 15,000 repeats of a crystallisation process before he finally produced a sample of pure thulium.

an incredible 15,000 times until the spectrum of the compound changed no more. Only then was he satisfied that he had produced a pure sample of the compound.

Portable X-rays

Around 50 tonnes of thulium are mined and separated each year, it is found primarily in its oxide as the ^{169}Tm isotope. If this is placed inside a nuclear reactor where it can capture a neutron, it can form the unstable ^{170}Tm isotope, with a half life of 128 days.

When ^{170}Tm decays to its stable neighbour ytterbium (^{170}Yb), it emits X-rays and has found use as a portable source of the radiation. Sources containing ^{170}Tm have a useable life of around one year, after which roughly 13% of the original sample remains. Because it decays to a stable isotope these sources are relatively safe to handle, requiring the cover of a simple lead cup. It is one of the four most popular medical isotopes for radiotherapy, finding use in dentists' surgeries.

Ytterbium

Yb

Ytterbium

70

Atomic number	70
Atomic weight	173.045
Abundance	3.2 mg/kg
Radius	175pm
Melting point	824°C
Boiling point	1196°C
Configuration	(Xe) 4f14 6s2
Discovered	1878 – J.C.G. de Marignac

Extracted from impure 'erbia' componds in 1878 by Jean Charles Galissard de Marignac, ytterbium was named, as other elements were, after the town of Ytterby, Sweden. Ytterbium shows some differences from other lanthanides.

I want it all

Where most lanthanides form compounds with a +3 oxidation state, ytterbium can also exist in an oxidation state of +2. These ytterbium (II) compounds are happy to donate electrons to find themselves in the +3 oxidation state; they are powerful reducing agents. They are so effective that they strip oxygen from water molecules, releasing hydrogen gas. Oxidation state options also make ytterbium a better catalyst than other lanthanides, particularly those used in organic chemistry.

Under pressure

Ytterbium is a soft, lustrous and silvery metal that tarnishes quickly in air because it is highly reactive. Under standard conditions it is a good conductor of electricity, but under pressure it deteriorates. This increase in electrical resistance makes it valuable in extreme pressure sensors that determine the magnitude of an earthquake, or the force felt near nuclear bomb explosions.

Differences discounted

Despite these unique uses, only around 50 tonnes of ytterbium are refined each year. This is mainly due to other lanthanides being so much cheaper that their poorer performance for the tasks at hand can be overlooked.

Large cracks appear in a road following an earthquake measuring 6.1 on the Richter scale on 20th March 2011 in Christchurch, New Zealand.

Lutetium

Lu
Lutetium
71

Atomic number	71
Atomic weight	174.9668
Abundance	0.8 mg/kg
Radius	175pm
Melting point	1652°C
Boiling point	3402°C
Configuration	(Xe) 4f14 5d1 6s2
Discovered	1907 – G. Urbain & C.A. von Welsbach & C. James

Lutetium is the rarest of the lanthanides, and consequently the last to have been discovered. It was, however, discovered by three different scientists in the same year.

Three's a crowd

In 1907 Georges Urbain in France, Carl Auer von Welsbach in Austria, and Charles James in the USA each prepared a sample of the metal oxide. Urbain, shortly followed by Welsbach, published early in the year before James. The International Commission on Atomic Weights, the chemistry authority at the time, chose to recognise Urbain's earliest publication as the discovery. He named element 71 from the Latin name for the city of Paris, Lutetia. Controversy remained as it was later found that Urbain's sample was impure, containing large amounts of ytterium, while Welsbach and James had indeed produced pure samples.

Cracking

Lutetium is used for very specialised tasks as its rarity and difficulty to extract make it an expensive commodity. As an oxide it catalyses the cracking of long carbon chains into smaller, more valuable, hydrocarbons. This process creates a number of alkene products which can then be turned into an array of plastics in a process called polymerisation.

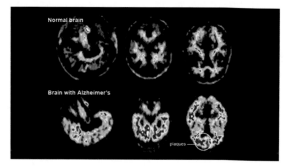

A positron emission tomography (PET) scan of an average brain and one of someone suffering from Alzheimer's disease, the most common form of dementia.

Viewing antimatter

Lutetium oxyorthosilicate (LSO) is a basis for electronic sensors that detect gamma radiation released in positron emission tomography (PET) scanners. Gamma radiation produced by positron-emitting isotopes in the body escapes the body to be read by the lutetium-containing sensors, and a 3D image of the body is reconstructed. This procedure images complex tissue such as the brain.

Atom Bomb

German physicists Otto Hahn, Fritz Strassman and Lise Meitner furthered earlier research, led by Enrico Fermi in Italy, by bombarding uranium with neutrons in the hope of discovering heavier elements. What they discovered changed the world forever.

Mushroom cloud produced by the detonation of XX-10 Priscilla, a 37-kiloton fission atomic bomb, on 24th June 1957, part of Operation Plumbbob. The bomb was suspended from a balloon around 200 metres above the Nevada desert before detonation. Many of the trans-uranium elements were first produced in trace amounts in explosion like this.

In early 1934 Meitner fled Germany as the persecution of Jewish citizens intensified, and settled in Stockholm, Sweden. Here she continued correspondence with Hahn and Strassman, who in 1938 discovered the process of nuclear fission. Meitner was able to theoretically explain the results seen by her male colleagues but was herself ignored by the Nobel committee, who awarded the 1944 chemistry prize to Hahn alone.

Nuclear fission

Meitner explained that nuclear fission, as we know it today, is the splitting apart of a heavy atomic nucleus into multiple smaller nuclei. The process releases large amounts of energy from a tiny difference in mass of these daughter products. Einstein informed us that the interchange between mass (m) and energy (E) is related by the speed of light in a vacuum (c). The famous $E=mc^2$ equation explains that just a tiny mass produces a huge amount of energy as the speed of light squared is a massive 9×10^{16} m2/s2!

Inducing

The process can occur naturally and spontaneously, or, as Hahn and Strassman observed, artificially and induced by firing a neutron into a heavy nucleus. Uranium compounds, bombarded by neutrons, were shown in their experiment to contain atoms of much lighter barium. This ability to control fission and the consequent release of vast amounts of energy made its use as a weapon theoretically possible. After declaration of war in 1939 both axis and allies accelerated research into use of this newly found science.

Developing the bomb

For around a year and a half, British research in the field outpaced that of the US. By mid-1942, however, it was clear that Britain could not stretch its already tight budget to cover the vast industrial costs required to make such weapons. Britain and the US agreed to collaborate in developing the bomb, without informing the allied USSR. Soon the US had ploughed so much money into the now-named Manhattan Project that it had become the largest industrial project ever undertaken. It was from this enterprise that atomic weapons were first realised and then used in a theatre of war.

Explosive material

Isotopes of heavy elements which undergo induced fission are said to be fissile. It was discovered that ^{235}U or ^{239}Pu were the perfect fissile candidates from which to build a bomb. When triggered by a release of neutrons, a large amount of energy is released as heavy nuclei split apart into lighter daughter nuclei and more neutrons. The neutrons released go on to induce fission of yet more heavy atoms, and the process continues. This chain reaction releases vast amounts of energy in a short amount of time, causing an explosion.

Building up

Release of many neutrons not only induces fission but also the possibility that a nucleus might capture neutrons. If this happens then neutron-rich isotopes can turn into new, heavier than uranium, trans-uranium elements through beta decay. Beta decay of a neutron in a nucleus spits an electron from the atom while also creating a proton; as the number of protons has changed we have a new element. It is this process that forms the lighter trans-uranium elements, and it is in the fallout of nuclear bombs that they were first man-made.

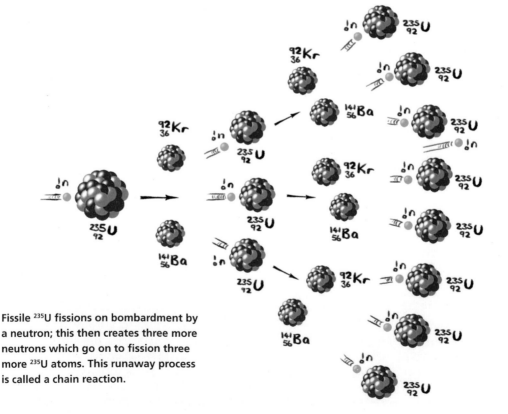

Fissile ^{235}U fissions on bombardment by a neutron; this then creates three more neutrons which go on to fission three more ^{235}U atoms. This runaway process is called a chain reaction.

Atom Smashers

Magnets can be pushed or pulled by a magnetic field; particles with an electric charge can be pushed or pulled by an electric field. In changing the electric fields experienced, man has built an array of machines that accelerate subatomic and ionised atoms to near light speed.

Spirals, circles and lines

Cyclotrons accelerate particles across a gap between two semi-circular cavities surrounded by a fixed magnetic field. The radius of a charged particle's circular movement increases as it speeds up, until the particle exits the machine at the edge of a cavity. Higher energies can be achieved using synchrotrons which tune the frequency of the accelerating electric field so the charged particles continually receive a push. As the particles then accelerate, the magnetic fields need to change so that they keep the particles travelling in a ring of the same radius. Linear accelerators are one-shot machines which use a similar technology to synchrotrons. They increase their push through changing electric fields but do not recycle uncollided particles, giving them just one chance of taking part in a collision.

Hitting the target

These machines are used to accelerate ionised atoms, which have had an electron or more removed. These ions are then slammed into stationary (not moving) targets which contain atoms of another, usually heavier, element. If enough energy is supplied then the nuclei of the ion and the stationary atoms will get close enough together that they have a chance to fuse together to form a single larger nucleus. If this happens then you have created a new element. In choosing the ionised atoms and stationary target carefully you can maximise chances of creating certain elements. This method is the one used to generate most trans-uranium elements (heavier than uranium).

A brief history

It all began with Ernest Lawrence's improved cyclotron design in the 1950s. It continued with synchrotrons which have today grown to the size of the 27km-circumference Large Hadron Collider at CERN in Switzerland, which discovered the Higgs boson in 2012. Today technology has advanced to the extent that linear machines just tens of metres long are sufficient to create the heaviest elements on the periodic table. These are responsible for the discovery of the last few elements up to 118 on the table.

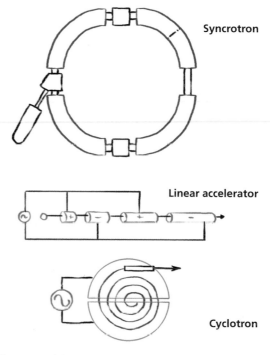

Syncrotron

Linear accelerator

Cyclotron

Different machines are used to accelerate electrically charged ionised atoms to synthesise new heavy elements.

Actinium

Actinium

89

Atomic number	89
Atomic weight	(227)
Abundance	5.5x10-10 mg/kg
Radius	195pm
Melting point	1050°C
Boiling point	3198°C
Configuration	(Rn) 6d1 7s2
Discovered	1899 – A.-L. Debierne

While sifting through the dregs left by Pierre and Marie Curie after their discovery of radium, Frenchman André-Louis Debierne discovered a new element. He announced the discovery in 1899, describing the element as similar to titanium, and later thorium.

Naming and claiming

Its high radioactivity led Debierne to name the element 'actinium' from the Greek for ray, aktinos, because of the rays of radiation it emitted. Three years later the German chemist Friedrich Oskar Giesel discovered the element in a similar experiment. He described it as similar to lanthanum and called the discovery 'emanium' after the radiation it emanated.

After three separate determinations of the half lives of each of the materials it was decided that Debierne's discovery was longer lived. Tradition therefore stated that the right to name the element was his, so actinium stuck.

Glowing

The cartoon idea that radioactive things glow is generally not true, but actinium is an exception. It spits out high-energy alpha particles which are fantastic at stripping electrons from surrounding atoms. When the electrons recombine with an atom they emit energy in the form of light.

Box containing vials of an early 20th-century preparation containing the highly radioactive isotopes of selenium, radium and actinium. At the time, radiation was viewed as the ultimate health cure, but today we know it to be a dangerous carcinogen.

Actinium is so alpha radioactive that a sample of the element is constantly surrounded by a blue halo.

Not many uses

When mixed with beryllium the 277Ac isotope creates a source of neutron particles. Such neutron sources are used in machines which search for water, kill cancer cells and scan cargo for bombs. The 225Ac isotope finds use in targeted alpha therapy where it delivers large amounts of dangerous radiation to kill cancer cells.

Thorium

Non-nuclear uses and a future fuel

Thorium
90

Atomic number	90
Atomic weight	232.0377
Abundance	9.6 mg/kg
Radius	180pm
Melting point	1842°C
Boiling point	4788°C
Configuration	(Rn) 6d2 7s2
Discovered	1829 – J. Berzelius

After a false start in 1815, when his discovery turned out to be yttrium phosphate, Jöns Jacob Berzelius eventually discovered thorium in 1828. Named after the Scandinavian god of thunder, thorium was soon to light up the world after dusk.

Glowing white

In 1891 Austrian chemist Auer von Welsbach was searching for a material that could withstand the intense heat of a gas flame. After investigating oxides of magnesium, yttrium and lanthanum, he discovered that thorium had the highest melting point of any known oxide: over 3300°C. When placed in a gas flame, the heated thorium oxide glowed with an intense white light, of much higher quality than that of the other oxides. Its properties led to thorium oxide being fabricated into delicate gas mantles, which for the first time provided the world with vast amounts of light after dark.

Thorium-containing mantles provide the intense white light given out by gas lamps.

All by myself

Although Berzelius did not know it at the time, the ^{132}Th isotope, that makes up pretty much all of the thorium on Earth, is radioactive. Only trace amounts of the other six naturally occurring isotopes exist at any one time as part of a chain of radioactive decay. The time in which it takes half of a collection of ^{132}Th atoms to radioactively decay is longer than the current estimate of the age of the universe, at 14.05 billion years.

When an atom does randomly change it is via alpha decay, an emission equivalent to a nucleus of helium. These alpha particles are highly electrically charged and very large so do not travel far through material, and are easily stopped by a piece of paper – or the glass surrounding a gas lamp.

Non-nuclear needs

In addition to gas mantles, thorium is mixed with tungsten to make arc-welding electrodes. These direct huge electric currents and associated heat to fuse metals together. The thorium improves the high temperature strength as it increases the crystal size of the tungsten metal. Thoriated gas mantles, purchased from a camping store, or thoriated tungsten welding electrodes, from a hardware store, are both cheap and safe forms of radiation used in classrooms worldwide.

Thorium is also added to glass to produce high-end camera and telescope lenses. Thoriated glass is able to bend and focus light efficiently and with less aberration (spreading of different colours of light) than other rivals.

Future fuel

Uranium, and heavier trans-uranium, elements produce nuclear waste from used fuel that remains dangerous to life for thousands of years. Thorium, on the other hand, offers an alternative, with waste that is safe within under 100 years. When bombarded with neutrons, and after a little radioactive decay, non-fissile ^{232}Th produces fissile uranium ^{233}U. When this isotope is mixed with molten fluoride salts it can be used as a nuclear fuel in molten salt reactors. The core of the reactor is surrounded by more ^{232}Th which absorbs neutrons emitted by the ^{233}U to continually breed more fuel to continue running.

Proponents are campaigning for funding to restart research into the technologies required, after it was ignored by most of the western world in the 1970s. Thorium is much more abundant than uranium and is as common as lead. Such vast amounts could supply our demand for energy long after fossil and uranium fuels run out. Another major argument is that thorium reactors do not breed fissile materials that can be weaponised. Despite these potential benefits, a 2011 study showed there is little chance that thorium will replace ^{235}U or ^{239}Pu as nuclear fuel of choice.

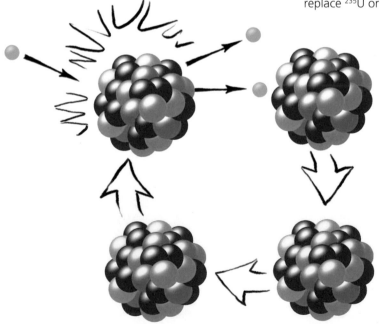

The nuclear fuel cycle in which the abundant ^{232}Th isotope (top right) absorbs a neutron from uranium fission of ^{233}U to become ^{233}Th. This then decays quickly to ^{233}Pa, which over a month decays to form ^{233}U, an ideal fuel. ^{233}U decays, releasing energy (white flash) and neutrons (blue spheres) to continue the cycle.

Protactinium

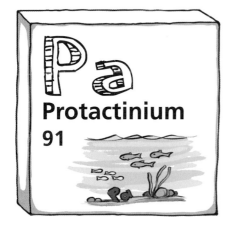

Pa

Protactinium

91

Atomic number	91
Atomic weight	231.0359
Abundance	1.4x10-6 mg/kg
Radius	180pm
Melting point	1568°C
Boiling point	4027°C
Configuration	(Rn) 5f2 6d1 7s2
Discovered	1913 – O.H.Göhring & K. Fajans

A gap in his 1871 table showed Mendeleev's prediction of an element that would lie between thorium and uranium. As the periodic table was constructed at the time, it was predicted to be a member of group 5 and have similar chemical properties to tantalum.

Uranium-X

In 1899, under the assumption of chemical similarity to tantalum, English scientist Sir William Crookes was able to crystallise a small sample of a highly radioactive material from uranium ore. Although the amount was too small to identify the elements present using spectroscopy, it was radioactive enough to expose photographic plate in just hours. He named the unknown substance "UrX" (uranium-X).

Brevium

In Germany in 1913 Kasimir Fajans and Otto Göhring used a similar method to extract and identify a sample as a new element. With a half life of just 6.2 hours the pair named their discovery 'brevium' as its existence was brief and fleeting. Just over four years later, Otto Hahn and Lise Meitner used a different method to extract what turned out to be the same element from uranium pitchblende. We now know this to be the most stable isotope of protactinium, ^{231}Pa, which has a half life of 35,000 years.

Meitner and Hahn, as discoverers of the longest-lived isotope, were granted the honour of naming the element. They named it protoactinium as it formed actinium when it decayed. This was later simplified to protactinium in 1949 by the IUPAC when they recognised Hahn and Meitner as the discoverers.

Protactinium is used to age the sediment found at the bottom of the ocean. Looking at the ratio of radioactivity coming from thorium and protactinium allows geologists to gauge how long the sea floor has lain undisturbed.

Uranium

Common and controversial

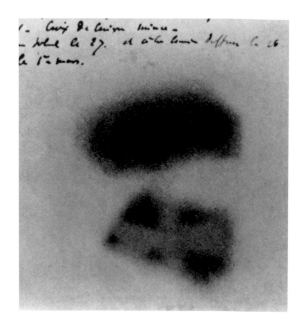

Atomic number	92
Atomic weight	238.029
Abundance	2.7 mg/kg
Radius	175pm
Melting point	1132°C
Boiling point	4131°C
Configuration	(Rn) 5f3 6d1 7s2
Discovered	1789 – M.H. Klaproth

Uranium symbol U, element 92.

Uranium is the heaviest naturally occurring element, thanks to its long-lived ^{238}U isotope with a half life comparable to the age of the Earth at 4.5 billion years. It is 40 times as abundant as silver, and is a hard white metal. It takes its name from the planet Uranus, which was discovered just a few years before it was.

Foggy fortune

Uranium is central to one of science's most famous accidental discoveries. French physicist Antoine Henri Becquerel tidied away his lab one evening and in one drawer, he placed some uranium salts on top of an unexposed photographic plate. When he returned the next morning he found that the plate had become 'fogged'. The conclusion was that some unseen rays of energy were being

This accidental photograph led to the discovery of radioactivity by Antoine Henri Becquerel in 1896. Dark patches show where the photographic plate, although left in a dark drawer, was exposed to some unknown radiation emanating from uranium salts. He won the Nobel Prize in physics for this discovery in 1903.

emitted by the salt: the first evidence of radioactivity.

Energy from the nucleus

Uranium is a mix of two isotopes: 99.3% ^{238}U and just 0.7% ^{235}U, which is of greatest interest for nuclear science. ^{235}U is fissile: it can be split apart by low-energy neutrons, in the process producing more neutrons. This allows it to sustain a chain reaction, which is essential for nuclear reactors and also bombs (see Atomic Bomb). The uranium used in nuclear reactors, usually in oxide form, is enriched to contain around 3% ^{235}U so that the nuclear reaction can continue.

Enrichment of uranium is a difficult process, requiring the separation of two very heavy isotopes with identical chemistry. The most widely used method of separation today is that of a gas centrifuge, which separates the compound uranium hexafluoride UF_6. In much the same way as heavy fruit within lighter cereal will end up at the bottom of a shaken packet due to the effect of gravity, the heavier $^{238}UF_6$ sinks below the $^{235}UF_6$ when spun around at high speed.

Non-nuclear

The many different oxidation states of uranium result in a variety of vivid colours of compound. Used to stain glass since the time of the Romans, they also stain wood, dye leather, and glaze ceramics today.

Depleted uranium contains less ^{235}U than it would naturally, around 0.2%, and is subsequently 40% lower in radioactivity. As it is a dense and heavy metal it is used to keep ships upright and aircraft balanced, acting as ballast and counterweights. It is also used in armour-piercing weapons ammunition as well as in the armour itself.

Ammunition specialist with depleted uranium-containing, 105mm armour-piercing rounds used by M-1 Abrams battle tank.

Neptunium

Atomic number	93
Atomic weight	(237)
Abundance	3x10-12 mg/kg
Radius	175pm
Melting point	644°C
Boiling point	4000°C
Configuration	(Rn) 5f4 6d1 7s2
Discovered	1940 – E.M. McMillan & P.H. Abelson

Np
Neptunium
93

In 1877, German chemist R. Hermann found what he thought to be a new element hiding in the mineral tantalite, which he named neptunium after the also elusive planet Neptune. In 1886 a second German chemist, Clemens Winkler, discovered another element and his initial thought for a name was neptunium. After finding out the name had been taken, he was resigned to naming it germanium after his country.

Taking its place

Later in the century, Hermann's 'element' was determined to be in fact an alloy of tantalum and niobium, but by this time 'germanium' was accepted for Winkler's element. So when, in 1940, the American duo Edwin McMillan and Philip Abelson discovered element 93, it took its rightful place between uranium and plutonium. The physicists created the element by bombarding uranium with neutrons inside a cyclotron particle accelerator. Although they did not know it at the time, their openly published paper was to show the world how to overcome one of the biggest hurdles in creating an atomic bomb.

Making other things

McMillan and Abelson found that ^{238}U captures a slow-moving neutron to form ^{239}U which decays, through beta decay (see Technetium), to produce

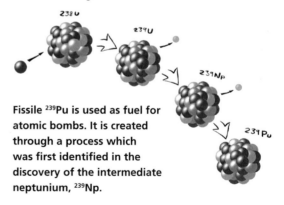

Fissile ^{239}Pu is used as fuel for atomic bombs. It is created through a process which was first identified in the discovery of the intermediate neptunium, ^{239}Np.

^{239}Np. This isotope is short lived: in just 2.4 days, half of it will have decayed through beta decay to form ^{239}Pu. This isotope of plutonium is fissile and perfect for use in atomic bombs. Through a similar process the isotope ^{237}Np is used to make ^{238}Pu, which is used to make nuclear batteries. Neptunium's only use is as a detector of high-energy neutrons, which split it apart.

Plutonium

At the heart of nuclear bombs

Atomic number	94
Atomic weight	(244)
Abundance	3x10-11 mg/kg
Radius	175pm
Melting point	639°C
Boiling point	3228°C
Configuration	(Rn) 5f6 7s2
Discovered	1940-41 – Seaborg, Wahl, Kennedy & McMillan

Our world changed forever in 1945 when the 'Little Boy' and 'Fat Man' atomic bombs exploded over the cities of Hiroshima and Nagasaki, Japan on 6th and 9th August. While the fissile ^{235}U isotope used in 'Little Boy' was hard to isolate from ^{238}U, the ^{239}Pu isotope can be generated inside a nuclear reactor.

Fat Man

Despite weighing over 4 tonnes, 'Fat Man' contained just 6.2kg of plutonium at its core. This sphere was below criticality, meaning that it would not simply begin the runaway chain reaction that led to an explosion. It was surrounded by explosives which, when detonated, compressed the ball of metal, bringing the atoms closer together, making it critical. To ensure the chain reaction the plutonium was also showered in neutrons from an initiator source, made from a mix of polonium and beryllium. This runaway chain reaction produced the energy released in the explosion.

Despite the fact that only around 1.2kg (20%) of the plutonium in the 'Fat Man' bomb actually detonated, 40,000 lives were lost in Nagasaki and up to 80% of the city was destroyed.

Waste product

The discovery in 1940 that ^{239}Pu could be made from bombarding non-fissile ^{238}U led many to design nuclear breeder reactors. These reactors resembled the nuclear reactors in power plants today, but focused almost solely on producing plutonium for bombs. Although plutonium has a similar chemistry, it is easier to separate from used nuclear fuel than the two isotopes of uranium themselves.

Around 1% of nuclear waste is plutonium, separated out today in the PUREX process; around 2 tonnes are extracted this way each year in civilian nuclear reactors. The greatest proliferation of plutonium was during the years of the Cold War, where the US and USSR refined 103 and 170 tonnes respectively. It is estimated today that there are around 500 tonnes of refined and weapons-grade plutonium on our planet.

Worse still

Perhaps the scariest use for plutonium is within fusion bombs. The process of nuclear fusion, creating heavier elements from lighter, releases around 1000 times more energy and consequent devastation. The energy from plutonium fission is used to heat and compress heavy isotopes of hydrogen to conditions not usually seen outside of a star. To date these 'hydrogen bombs' have only been tested and never actively used in warfare.

Pee and pu

The UPPu club was set up by scientists working at Los Alamos Lab, California, at the height of World War II. Those working on the Manhattan Project developing the bomb had to regularly handle plutonium, and there were inevitable accidents. As it is entirely man-made plutonium has no known biological role and if it gets into the body it lingers for a long time. The workers found that when they went to the toilet their pee would always contain trace amounts of plutonium; hence the name of the club: You-pee-Pu (UPPu). It is reported that one worker could detect small amounts of the radioactive metal 50 years after the war had

Composite image of NASA's Curiosity rover at the Rocknest site on the surface of Mars, near the base of Mount Sharp (upper right). Curiosity was kept warm during the cold Martian nights by a plutonium battery.

ended. He was lucky, because plutonium is not only radioactive but also a toxic heavy metal; a fatal dose is just a few hundred micrograms.

Accident and recovery

During the war British radio chemist Alfie Maddox spilt the entirety of the UK's sample of plutonium, which was only about 10mg. He grabbed a saw and cut off the section of the desk he had spilt it on, burnt it and then extracted the plutonium carefully from the ashes. He managed to reclaim 9.5mg of the original sample to continue his research.

Positive plutonium

The radioactivity of plutonium is also put to the positive endeavour of exploring our solar system and beyond. Plutonium batteries, where the radioactivity provides both electricity and warmth, have been used on a number of different space missions. The Curiosity rover currently surveying the planet Mars and the New Horizons Project, aptly destined for Pluto, each owe their energy to plutonium.

Pu profile

Plutonium is very interesting as a material. It is very difficult to cut and machine because it exists in several allotropes simultaneously, each with a different hardness. If left for a very long time it breaks apart as the helium gas produced from alpha decay makes the metal porous. It has a colourful chemistry similar to uranium.

The secret life of the periodic table

Americium

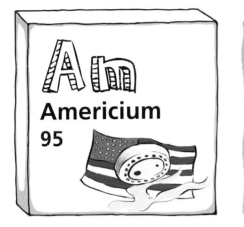

Am
Americium
95

Atomic number	95
Atomic weight	(243)
Abundance	0 mg/kg
Radius	175pm
Melting point	1176°C
Boiling point	2607°C
Configuration	(Rn) 5f7 7s2
Discovered	1944 – Seaborg, James, Morgan & Ghiorso

Element 95 was discovered in 1944 by Glenn T. Seaborg and his colleagues working as part of the Manhattan Project at the University of Chicago. Their discovery was not revealed until November 1945, however, as they were sworn to secrecy. The allies did not want anyone knowing exactly how far they had come in the field of nuclear physics.

All hell

The team that first isolated the element had to go through an arduous process which involved many stages of purifying nuclear waste, including burning, dissolving in acid, and being bombarded by particles in a cyclotron accelerator. Because of this, they jokingly suggested calling the element pandemonium, a word taken from the Greek for all hell. Instead it appropriately took its place under europium, named after the continent in which it was found: America.

Many processes are needed to isolate americium from radioactive waste, keeping scientists very busy.

Is it smoky?

Americium is the only trans-uranium element that all of us are likely to have in our house. There is a fraction of a microgram of the isotope [241]Am in smoke detectors where it sits decaying into alpha radiation. Alpha particles can strip electrons from atoms as they pass through the air, which if collected results in a tiny electric charge flowing. While the electric charge flows from this decay no alarm sounds. If smoke particles enter the detector, however, they readily absorb the alpha radiation. As this stops the alpha particles removing electrons from the surrounding air, it leads to a drop in the electric current which then sounds the alarm.

Alternate?

Because strict regulations surround plutonium, americium is being considered by the European Space Agency (ESA) as an alternative source of energy in nuclear batteries. It is extracted from nuclear waste, containing just 1 gram in every tonne.

Curium

Cm

Curium

96

Atomic number	96
Atomic weight	(247)
Abundance	0 mg/kg
Radius	no data
Melting point	1340°C
Boiling point	3110°C
Configuration	(Rn) 5f7 6d1 7s2
Discovered	1944 – Seaborg, James & Ghiorso

Marie and Pierre Curie take their rightful place in the roll call of scientists, giving their name to the radioactive element 96.

Dangerous energy

Element 96 is extremely radioactive, with a short half life for all isotopes, so consequently releases huge amounts of energy. While energy from americium and plutonium has been harnessed to produce electricity, the high-energy gamma radiation emitted by curium makes it impractical to handle. Its only real use has been to supply alpha particles as probes to analyse Martian soil. The Mars exploration rovers have taken a little curium with them in technology known as alpha particle spectrometers, to determine what the red planet is made from.

The Curies

Maria Skłodowska moved to France from Russian-controlled Poland and gained a degree in physics in Paris in 1893. While searching for laboratory space for her research she met Pierre Curie. The two found an instant connection in their interests and a short time later Pierre proposed marriage.

After their marriage in 1895 Marie, as she was called by French friends, was inspired to research uranium rays, recently discovered by

Henri Becquerel. Using Pierre's electrometer she demonstrated that uranium radiation allowed the surrounding air to conduct electricity. Marie hypothesised that whatever was being emitted must come from within the atom of the element.

Discoveries

In early 1898 Marie suggested that the ores of pitchblende and chalcolite contained new radioactive elements. With Pierre now helping, the couple processed tonnes of pitchblende and published the discovery of polonium in July and radium in December.

Pierre and Marie Curie photographed in 1895, the year they married.

Berkelium/Californium

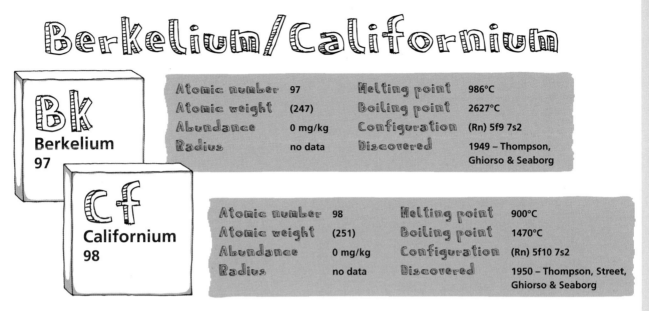

Atomic number	97	Melting point	986°C	
Atomic weight	(247)	Boiling point	2627°C	
Abundance	0 mg/kg	Configuration	(Rn) 5f9 7s2	
Radius	no data	Discovered	1949 – Thompson, Ghiorso & Seaborg	

Bk
Berkelium
97

Cf
Californium
98

Atomic number	98	Melting point	900°C	
Atomic weight	(251)	Boiling point	1470°C	
Abundance	0 mg/kg	Configuration	(Rn) 5f10 7s2	
Radius	no data	Discovered	1950 – Thompson, Street, Ghiorso & Seaborg	

Present in tiny amounts in the fallout of nuclear bomb explosions, berkelium was first synthesised in measureable amounts using a particle accelerator. With just three-billionths of a gram, Glenn T. Seaborg, Albert Ghiorso and Stanley G. Thompson could only identify element 97 through the spectrum of light it emitted.

As with americium and curium before, the naming of this new actinide followed the lanthanides above. Terbium was named after its place of discovery, Ytterby, so it seemed fitting that element 97 be named berkelium as it was discovered in University of California, Berkeley. The tradition did not continue beyond this element; element 98 was discovered almost simultaneously and simply named californium.

Californium, also discovered at Berkeley Lab, was named after the western US state in which it was discovered. It is used primarily because of its desire to emit neutrons, with each microgram of newly generated ^{252}Cf emitting over 2.3 million neutrons each second.

In looking at how neutrons scatter when they pass through a material you can determine its composition. Californium supplies the neutrons used in machines that use this principle for oil, water and precious metal prospecting. The similar process of neutron tomography scans aeroplanes in airports around the world, searching for possible weakness or fatigue in their metal structure.

Neutrons are also required for igniting the chain reaction of fissile material in a nuclear reactor and californium provides this spark. It has also been used as a target which, when bombarded with a lighter element, created the final row elements 103 and 118.

A farmer uses a californium-based neutron probe to determine the moisture content of soil.

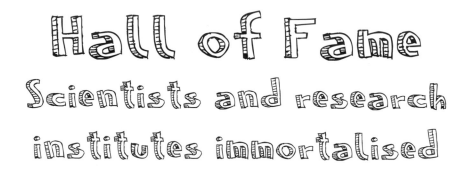

Hall of Fame
Scientists and research institutes immortalised

Although primarily the domain of chemistry, we owe our understanding and the discovery of the heavy elements to the science of physics. The Solvay Conferences in the early 20th century, founded by Belgian industrialist Ernest Solvay, collected together the finest scientific minds of the time. Here they discussed each great scientific advance of the age, with every conference having a unique focus. At the most famous conference in 1927 on the topic of 'Electrons and Photons' the foundations were completed for newly formulated quantum theory.

In 1933 the Solvay Conference focused on the 'Structure and Properties of the Atomic Nucleus'. Among those attending were no less than six scientists who later gave their names to elements on the periodic table. Albert Einstein (einsteinium) was missing from the photograph but sitting around the table we find Niels Bohr (bohrium, 107), Marie Curie (curium, 96), Ernest

This photograph taken at the 1933 Solvay conference contains no less than six of the scientists acknowledged in the periodic table with elements named after them.

The secret life of the periodic table

Rutherford (rutherfordium, 104), and Lise Meitner (meitnerium, 109). Standing behind them we also see Enrico Fermi (fermium, 100) and Ernest Lawrence (lawrencium, 103). This photo probably shows the largest number of scientists with elements named after them collected together in one room.

Super-heavy and man-made

The final super-heavy elements have only been produced in minute amounts, have no practical use, and their chemistry can only be theoretically predicted. They are all man-made inside particle accelerators where ions speed up to fractions of the speed of light before being slammed into targets. With patience, and a bit of luck, the scientists catch a glimpse of just a handful of atoms of these new elements. Their nuclei are so unstable that they last only fleetingly before decaying into lighter elements once more, leaving little or no time for any chemical reactions to be observed. However brief, observations of these super-heavy elements have shown the power of quantum physics in predicting their properties. A precedent had been set with lighter elements to

allow the newly discovered elements to be named after places, institutions and famous scientists. Because of the lack of properties to speak of, in this section we mainly focus upon the scientists and institutes which rightly give their names to these elements, with a few tit-bits about the elements themselves.

Es
Einsteinium
99

Einsteinium

As with many of the super-heavy elements, element 99 was first made on Earth in the explosion of a thermonuclear H-bomb, but has been made in particle accelerators ever since. The lighter super-heavy elements like einsteinium are only really used to synthesise other super-heavy elements; as a target upon which lighter ions are slammed at high energy. Einstein earns his immortalisation as an element thanks to his contribution to the early understanding of subatomic particles and development of quantum theory (see Quantum Atom). Also, his special theory of relativity allows us to explain the change in chemical behaviour that occurs as atoms become large. The inner electrons, in the $1s$ shell, in large atoms travel at a high proportion of the speed of light and one cannot ignore the effects of relativity. The result is a change in the energy levels of electrons around the nucleus, which leads to differences in chemistry and the light they absorb and emit. It is relativistic effects that make gold yellow, cobalt blue and copper orange. Einsteinium does live long enough to form chemical compounds. Einsteinium nitrate is formed when the metal is reacted with nitric acid and einsteinium oxide is formed by burning this nitrate. Einsteinium halides such as einsteinium chloride or fluoride have also been formed. It is fitting that these compounds show the predictive power of the quantum theory Einstein first help to build.

Albert Einstein, probably the most recognisable face of science, contributed so much to our understanding of the atom.

Fermium

Italian physicist Enrico Fermi has been called the 'architect of the nuclear age'. He was the first person to observe artificial radioactivity after bombarding thorium and uranium with neutrons, although at the time he thought he had discovered new trans-uranium elements. In 1938 news came from Germany of an experiment by Otto Hahn and Fritz Strassmann that showed the uranium forming the lighter element barium, among other things, when bombarded by neutrons. This was correctly identified as the uranium atom splitting apart into lighter elements by exiled Jewish German scientists Lise Meitner and Otto Frisch: the process is known today as fission. Fermi moved to the US shortly after fascists took control of Europe during the Second World War and continued his research, joining the Manhattan Project, the US effort to build the first atomic bomb. Along with other US and European refugee physicists he built the world's first nuclear reactor in a squash court, under the football stadium at the University of Chicago. His research greatly accelerated the allies' development of the atomic bomb and it was within the fallout from a such a bomb that element 100 was first made on Earth.

A Nobel Prize medal is adorned with the face of chemist Alfred Nobel. The name of the element is actually associated with the Nobel Institute where a number of elements were discovered, not the man himself.

Mendelevium

Pages have been dedicated already to Dmitri Mendeleev (see Mendeleev and the Modern Table), the founding father of the modern periodic table. One myth that hangs around the great man comes from his time working for the Russian government, where he defined standard weights and measures. He is said to have set the standard strength of vodka as 40% ethanol in water. Element 101 was first made in Lawrence Berkeley National Lab in the US in 1955 by bombarding lighter einsteinium with alpha particles. Today the element is made primarily by slamming ions of argon into a bismuth target.

Nobelium

Alfred Nobel was a Swedish industrial chemist who made his fortune making and selling explosives. Early explosives were unstable carbon and nitrogen compounds, which easily ignited and initiated runaway chain reactions. This rapidly turned small amounts of solid into large amounts of hot gases and often blew off a limb or two. Nobel discovered that spreading the compounds onto an inert, unreactive, substance meant they would only ignite if a short sharp explosion, called a detonator, was used. This allowed safe(r) handing of explosives which led to expansion of the railways and advances in warfare, and Nobel made lots of money. During his lifetime Nobel set up a research institution in Stockholm, and it was here in 1957 that a team became the first to claim discovery of element 102. They had produced it by bombarding a curium target

The secret life of the periodic table

with the radioactive isotope carbon-13. Years of controversy followed, as American and Russian groups were not able to repeat the experiment. Despite the Russian Joint Institute for Nuclear Research (JINR) Dubna group producing the first concrete evidence of the element, the IUPAC decided the Nobel group be granted the naming right. Upon Nobel's death a foundation was set up to award prizes which recognise great advances in all areas of science and also world peace. The awarding of these have been as controversial as element 102, especially the way in which the contribution of many great women, such as Lise Meitner, has been ignored entirely.

Lawrencium

Ernest Lawrence invented the cyclotron particle accelerator and built and used a 60-inch version of the machine at the University of California, Berkeley. This machine discovered new elements, but also hundreds of new isotopes of elements already known. The lab was later named the Lawrence Berkeley National Lab in Lawrence's honour and it was here that the claim for element 103 first came in 1961. Tracks were seen by Albert Ghiorso and team after bombarding a target of differing californium isotopes with boron isotopes. With a claim also from Dubna in Russia, where americium was bombarded with oxygen, the naming of the element was debated across the iron curtain throughout the Cold War. Although the IUPAC gave naming rights to the American group, in 1997 it rightly recognised the Russian group from Dubna as co-discoverers.

Rutherfordium

New Zealand-born Ernest Rutherford played a pivotal role in our understanding of the atom, discovering the nucleus and proton (see Atomic Physics). Lord Rutherford had a deep booming voice which often affected the sensitive experiments conducted by his students. It annoyed them so much that they put up signs asking everyone to talk softly in the laboratory.

Rutherford had a very different work ethic to most academics and imposed this upon his students. While most academics worked their students at all hours, Rutherford would not allow his students to work in the lab after 6pm. When this time came he ensured that work would stop by asking his chief lab technician to sweep through the lab switching off all equipment. There must have been something worthy in this behaviour because a record eleven of his students continued in their career to win a Nobel Prize. Rutherfordium itself was discovered by the team at the JINR in Dubna, Russia in 1964 after neon ions were collided into an americium target.

New Zealander Ernest Rutherford is one of the giants in the field of atomic and subatomic physics and trained some of the best scientists of the proceeding generation.

Dubnium

The JINR in Dubna is something of an element factory. The scientists there have either discovered, co-discovered, or confirmed every man-made super-heavy element on the periodic table today. During the Cold War there were a number of debates regarding the discovery of elements. The JINR Dubna group in Russia published a result in 1968, seen in the collision of neon ions with americium, while a group at the University of California, Berkeley claimed discovery of the same element in the same year after colliding nitrogen ions with californium. The debate over discovery and naming rights came to a head in the 1990s when the naming of the super-heavy elements was eventually agreed by the IUPAC after consultation with American, Russian and European working groups; element 105 was officially given the name dubnium in 1997.

Seaborgium

You could have sent a letter to Glenn T. Seaborg during his time working at the Lawrence Berkeley National Laboratory in California, using only the names or chemical symbols of elements:

Seaborgium	Sg
Lawrencium, Berkelium	Lr, Bk
Californium	Cf
Americium	Am

Of the five it is, ironically, only the element named after him that Seaborg did not have a direct hand in discovering. It was jointly discovered by researchers at the JINR in Dubna, and another group of physicists at Lawrence Berkeley Lab in 1974.

Originally the IUPAC recommended the element be named rutherfordium, after adopting the rule that no element could be named after a living scientist. When chemists argued that einsteinium was suggested during Einstein's lifetime, they re-examined their position. A reshuffle in 1997 saw elements 104–108 renamed: element 106 became known as seaborgium and rutherfordium was instead assigned to element 104.

Bohrium

This was the first element to be made under 'cold' nuclear fusion, in which the ions that are aimed at the target have relatively low energy compared with the creation of the other elements. In the case of bohrium it was low-energy chromium ions using a bismuth target. This process was pioneered by scientists at the JINR in Dubna,

Glenn T. Seaborg holds a balance which was used for the first weighing of plutonium, the element which he co-discovered.

where the first claim for discovery came in 1976. They originally named the element neilsbohrium, with symbol Ns. With doubt surrounding this claim, the IUPAC instead recognised a 1981 discovery by German scientists at the Gesellschaft für Schwerionenforschung (GSI) lab in Darmstadt. To acknowledge the pioneering work of the JINR scientists, the GSI group chose the same name they had originally suggested. Alongside the discovery acknowledgement, in 1992 the IUPAC also simplified the name of the element to bohrium and the symbol became Bh.

Hassium

Officially element 108 was discovered in 1984 at the GSI laboratory in Darmstadt. Translating as the Institute for Heavy Ion Research, the laboratory is located in the Hesse region of Germany, which gives this element its name. It was first made by slamming iron ions into a lead target. Although isotopes of this element have a half life of just a few seconds, this has been long enough to do some chemistry. Comparisons with osmium, directly above in the table, has shown possible interesting properties. Passing the few hassium atoms created through oxygen, scientists have made the compound hassium tetroxide. As this does not evaporate as easily as osmium tetroxide it suggests hassium might have a higher melting point than osmium. However, this all depends on the strength of interaction between hassium atoms which is unknown, as it would require tens of millions of atoms to be made.

Meitnerium

This element immortalises Lise Meitner who, along with Otto Hahn, discovered protactinium. The pair also played an essential part in the field of radioactivity, when in

Wilhelm Roentgen discovered X-rays in pioneering experiments with electromagnetism and is immortalised in the name of element 111.

1938 they discovered the fission of thorium and uranium from the new elements produced in their natural decay. Hahn was recognised for his part in this work and awarded the 1944 Nobel Prize in Chemistry; Meitner, however, was totally overlooked. Element 109 was officially discovered in 1982 at the GSI in Darmstadt after ions of iron were collided with a bismuth target. In 1994 the name was suggested to the IUPAC to finally recognise the work of Meitner. It is the only element on the table to be named solely after a non-mythical woman (curium is named after both Marie and Pierre).

Darmstadtium

The major heavy element laboratories around the globe each lend their name to one or more of the super-heavy elements. The GSI in Darmstadt, Germany, discovered six super-heavy elements: bohrium, hassium, meitnerium, darmstadtium, roentgenium and copernicium. Element 110 is named after the town in which the institute resides. It was made by slamming high-energy nickel ions into a lead target where the two occasionally fused.

Roentgenium

Element 111 was also discovered at the GSI, in 1994, and is named after the German Physicist Wilhelm Roentgen (Röntgen). Roentgen won the first Nobel Prize in Physics, when the awards began in 1901, for his work on X-rays. While experimenting with electromagnetism in 1895 he produced and detected X-rays for the first time. Roentgenium was formed by slamming nickel ions into a bismuth target.

Copernicium

Named after Polish astronomer Nicolaus Copernicus who first suggested it was the Earth that orbited the sun (heliocentrism) and not the other way around. This triggered the Copernican Revolution, a period of great advancement in our scientific understanding of nature.

When the name was first suggested there was widespread dislike for the suggested symbol of Cp. Organic chemists had for years been using Cp as shorthand for the Cyclopentadienyl (C_5H_4-)

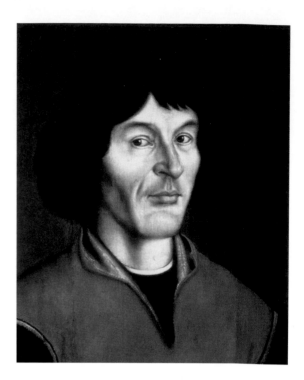

Nicolaus Copernicus (1473-1543) was a Polish astronomer who devised the Heliocentric model of the universe, where the Earth and the other planets orbit the sun.

ion, important in organometallic chemistry (see Zinc). The same symbol had also been used by German scientists as late as the 1950s as they referred to element 71 as cassiopium (Cp), now lutetium. As historic papers were digitised and searches automated, confusion would arise from these other uses. After a period of deliberation, the IUPAC decided to assign element 112 the symbol Cn. It was first synthesised at the GSI in 1996, and made by slamming zinc ions into a lead target.

Ununtrium

The heaviest elements are identified by just a few atoms at most. Living such a short time, confirmation that they have been created comes from a measurement of daughter

The secret life of the periodic table

elements produced in their chain of radioactive decay. Discovery of element 113 hinged upon clear identification of the well-known stage where dubnium decayed to form lawrencium. The first claim came in August 2003 from a US–Russian collaboration, between Livermore National Lab and JINR Dubna, but they were not able to identify this crucial stage with great enough confidence.

Discovery was instead awarded to the Japanese RIKEN group in December 2015 when their first claim from July 2003 was backed up by repeat experiments in April 2005 and August 2012. They created just one atom of element 113 each time, from trillions of zinc ions colliding with a bismuth target, but were able to confirm the decay chain of the element to be:

$$^{278}\text{Uut} \rightarrow {}^{274}\text{Rg} + \alpha \rightarrow {}^{270}\text{Mt} + \alpha \rightarrow {}^{266}\text{Bh} + \alpha \rightarrow$$
$$^{262}\text{Db} + \alpha \rightarrow {}^{258}\text{Lr} + \alpha \rightarrow {}^{254}\text{Md} + \alpha$$

At the time of writing the world is still waiting for the RIKEN group to announce their suggested name for the element. Once suggested it can take up to five months for the IUPAC committee to meet and approve any suggested name. It is currently assigned a IUPAC placeholder name: these are formed from the Latin numerals un for one, bi for two etc. Ununtrium is simply 'one-one-three-um' in Latin. All new elements that are yet to be named, or in fact discovered officially, first take such a place holder name. This element will be the first to be named by Japanese, or indeed any Asian, scientists.

Georgy Flyorov was the main proponent of the atomic bomb in the former USSR and was director of the nuclear reactions laboratory where a number of the trans-uranium elements have been discovered.

Flerovium

Flerovium
114

A single atom of element 114 was first thought to have been created in December 1998 at the JINR in Dubna, after calcium ions were slammed into a plutonium target. Although this exact result could not be repeated, other signals were seen in March of 1999 when the ^{244}Pu target was replaced with a lighter ^{242}Pu isotope.

Confirmation came in June 1999 when the JINR scientists repeated the 1998 experiment and saw results consistent with those from March.

The name of element 114 was officially adopted by the IUPAC on 30th May 2012; it was named after the founder of the JINR, Georgy Flyorov. It is a fitting memorial to a great scientist, alive when the discovery was made; Flyorov sadly passed away before the naming of element 114 was officially announced.

Ununpentium
Ununpentium
115

The discovery of element 115 is one of a collection rising from the fruitful partnership between Lawrence Livermore National Lab in the US and the JINR in Dubna, Russia. Calcium ions were fired by the JINR accelerator into an americium target supplied by Livermore Lab.

Livermore was first set up at the height of the Cold War to develop technologies used in nuclear warfare. It is perhaps ironic then that today it generates, purifies and supplies radioactive heavy elements to the JINR in Russia.

It is now known that element 115 decays via alpha decay to form ununtrium before following the same decay chain. The discovery was confirmed by GSI in Germany in August 2013. At time of writing, this element is awaiting a name from the scientists of the JINR and Livermore National Lab.

Livermorium
Livermorium
116

Element 116 is named after the Lawrence Livermore Lab but was discovered at JINR in Russia. Curium was made in nuclear reactors at Oakridge National Lab and prepared by Lawrence Livermore National Lab in the US before being flown to the JINR in Russia where calcium atoms slammed into it to produce element 116. The US–Russian collaboration is an example of how science can bridge political differences, and is in opposition to the original intention of the two labs during the Cold War.

Ununseptium
Ununseptium
117

This is another success of US–Russian collaboration, this time directly between Oakridge National Lab and the JINR. Oakridge supplied the berkelium target to the Dubna group, who then bombarded it with calcium ions. Ions of light elements like calcium are used because larger

Discovery of all super-heavy elements has taken place in labs based in the USA, Germany, Russia and Japan. Ununtrium is the first element to have been discovered in Japan and also the first element to have been officially discovered in Asia.

The secret life of the periodic table

Timeline of discovery for the trans-uranium elements

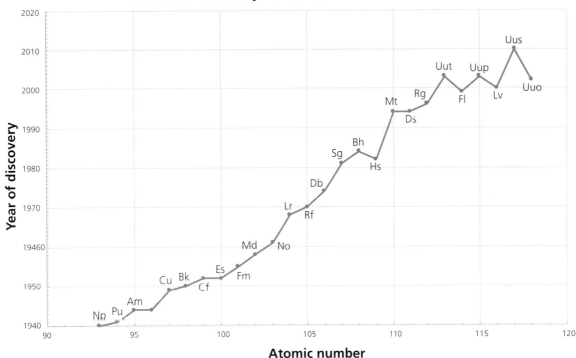

Year of discovery (y-axis): 2020, 2010, 2000, 1990, 1980, 1970, 19460, 1950, 1940

Atomic number (x-axis): 90, 95, 100, 105, 110, 115, 120

Data point labels: Np, Pu, Am, Cu, Bk, Cf, Es, Fm, Md, No, Lr, Rf, Db, Sg, Bh, Hs, Mt, Ds, Rg, Uut, Fl, Uup, Lv, Uus, Uuo

ions would be more likely to obliterate any target they hit and less likely to fuse with it to form new elements. This means that there may be a real physical limit to producing elements in this way.

As the discovery was announced in January 2010 it is the last of the elements currently on the periodic table to have been synthesised and therefore discovered. Odd-numbered elements are less stable than their even-numbered neighbours for reasons of symmetry. Even numbers of nucleons (protons and neutrons) allow a greater number of possible configurations which leads to a more stable lower energy nucleus.

Ununoctium
118

Ununoctium

We have finally reached the last element in the periodic table as it stands in 2016. It has a full electron shell but is likely to behave very differently to the other noble gases due to

This graph shows how as time went on scientists uncovered heavier and heavier elements.

relativistic effects and its sheer size. Announced in October 2006, it was a collaboration between Lawrence Livermore National Lab and the JINR in Dubna, which created three atoms of element 118: one in an initial experiment in 2002 and two more in 2005. They were produced after patiently smashing innumerable krypton ions into a lead target. To prove they had seen ununoctium, scientists had to observe its decay via alpha decay into flerovium and the decay chain that followed. To do this they first had to gain more evidence of element 116, measure its decay signatures, and then prove that the additional decays seen in 2002 and 2005 were due to element 118.

It worked, and along with elements 115 and 117, in collaboration with the Oakridge National Lab, the US–Russian team were granted naming rights by the IUPAC, who officially recognised the discoveries.

Future Elements

Armed with knowledge of geometry and astronomy, European explorers of the 15th century used the stars to navigate their boats around the globe in search of new lands. In the 21st century modern element hunters are doing something rather similar.

Electromagnetic force

An electromagnetic force repels particles of a similar electric charge or polarity of magnetic field. Bring a north pole of a magnet towards another north magnetic pole and they will push back. The same is true for electrically charged particles. The closer they are to each other the greater the push, but with sensitive enough equipment, the influence of the force could be felt no matter what the distance is between electric charges. The range over which an electromagnetic force can be exchanged is infinite. This means that each proton in an atomic nucleus, regardless of its location, pushes on every other proton. Those located on opposite sides push less than those side by side, but they still push. Add more and you increase the overall pushing force experienced by each proton.

Strong force

Protons remain tightly packed together in a nucleus despite their pushing thanks to the strong nuclear force. This force is attractive and is felt not only by protons but also neutrons, and it pulls them together. Unlike the opposing electromagnetic force, the strong force has a limited range and is only felt by directly neighbouring protons or neutrons. It is much stronger than the electromagnetic force and so

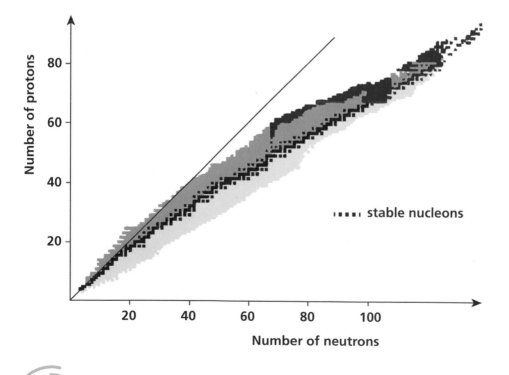

..... stable nucleons

Number of protons (y-axis)
Number of neutrons (x-axis)

Scientists are now setting sail from the shore of one island of stability in search of another. At the centre of this new land is element 122, thought to be stable enough that it would be around long enough to be detected.

it overcomes the desire of protons to move away from each other. However, as it only has a short range it does not add up to a greater and greater force as you add more protons or neutrons but instead remains constant.

Unstable

There therefore has to come a point where a nucleus has so many protons that the electromagnetic force has added enough to overcome the much larger but constant strong force. This nucleus is unstable and loses some protons and neutrons to form a lighter nucleus in which the strong force is more dominating. This is nuclear decay and it is usually done by ejecting a helium nucleus in alpha decay or heavier stable nuclei through fission.

The half life of an isotope is the average time taken for half of a collection of atoms to undergo decay. It is a direct measure of how unstable an isotope is; the smaller the half life the more unstable. Stable nuclei have infinite half lives. The increasing electromagnetic force and constant strong force means that the heavier an atom becomes the shorter its half life. The pattern shows an exponential decrease and so any new elements that are to be made would most likely have half

lives in the region of billionths of a second, which makes their identification very difficult indeed.

Necessary neutrons

Neutrons have zero electric charge and can be added to a nucleus without changing the identity of the element. They are there because they space protons apart a little, reducing the repulsive electromagnetic force, and contribute to the binding strong nuclear force. To cope with increasing outward pressure from larger numbers of protons pushing themselves apart, bigger proportions of neutrons are added to heavier elements.

These neutron-rich isotopes of elements are difficult to create from lighter nuclei that tend to have more of an even 50/50 split in proton and neutron numbers. The neutron deficient isotopes of the discovered heavy elements above 100 therefore show very short half lives close to the exponentially low values predicted.

Stability

As discussed when talking about oxygen, nuclei which have full nuclear shells of protons or neutrons are more stable than would otherwise be expected. These magic nuclei tend to show half lives greater than that predicted from the simple model described above. Nuclei with a full proton and full neutron shell are doubly magic and are many times more stable than those isotopes around them. To create heavy nuclei which can be observed, scientists are focusing on prediction of super-heavy doubly magic nuclei.

Predicting elements

Predicting stable heavier elements requires in-depth knowledge of the heaviest currently known. Particular focus has been given to flerovium after research in the 1960s suggested the isotope ^{298}Fl could be doubly magic and have a half life of minutes, not the microseconds, observed of current super-heavy isotopes. If it exists, it would be at the centre of an island of stability in a sea of unstable atoms.

Using current particle accelerator methods there is no way of synthesising ^{298}Fl as no combination of target and projectile atom can provide the massive 184 neutrons required. Methods suggested to provide the requisite neutrons involve the collision of man-made radioactive neutron rich isotopes or controlled nuclear explosions.

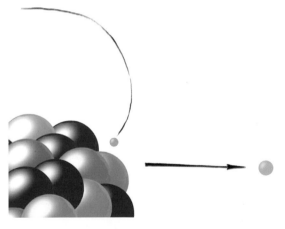

The inner electrons of super-heavy nuclei will reach the point where they will pair electrons and antimatter positrons just to lose energy. The electron will nosedive into the nucleus while the positron is ejected from the atoms.

Next element

Patterns in the half lives of neutron-deficient heavy isotopes created so far give the clearest indication of the road of discovery. Scientists have predicted the next nucleus with magic numbers of protons would be element 122, unbibium (Ubb). The isotope ^{306}Ubb would be double magic, but again it would need new methods to create such a neutron-rich atom. Nevertheless, labs around the world continue to attempt to create neutron-deficient isotopes of elements 120 and greater which may still live long enough for detection as they lie on the shores of yet another island of stability.

Feeling trapped and speeding up

It may seem that with improved equipment and patience one could go on forever adding elements to the periodic table, but there is a real physical limit to the size atoms can take. Larger atoms contain more protons and a more positive electrically charged nucleus. This greater charge attracts and binds orbiting electrons tighter. The quantum uncertainty principle states that if we have a good idea of where to find an electron, we have very little idea of how fast it is moving. The result is that the smaller the box you put an electron in, the faster it moves. Tightly bound electrons are kept in very small regions and so their speed is very great. Confine an electron enough and its movement rapidly approaches the speed of light.

Faster than light?

There reaches a point where the electron's speed has a possibility of exceeding the speed of light, but as this is impossible it instead finds ways of losing energy so that it travels more slowly. It achieves this by creating pairs of electrons and antimatter positrons. While the negative electrically charged electron nosedives into the atomic nucleus to be captured, the positive positron is accelerated out of the atom. This system is not physically stable and if created would last such a short amount of time that it could not practically be seen, and

The secret life of the periodic table

US Nobel Prize-winning physicist Richard Feynman predicted the largest possible atom from knowledge of how electrons and protons interact through the electromagnetic force.

therefore cannot exist. An analogue of such an atom has been shown to exist using the carbon allotrope graphene (see Carbon), showing that the instability does indeed result in atoms emitting positrons.

Feynmanium

This was first suggested by physicist Richard Feynman who related it to the fine structure constant of the electromagnetic force, which is in a way a measure of the resolution with which we can observe nature. This gave a naïve approximation that the breakdown would occur after element 137, dubbed by some as 'feynmanium'. Many followed up this suggestion and fleshed out the prediction with a more detailed approach. When all interactions between the subatomic particles are taken into consideration, breakdown is currently estimated to occur around element 173.

Conclusion

These pages have aimed to show the vast array of uses and rich stories surrounding every known type of atom in the Universe. We may have come a long way from the days of alchemy, but there is still much to learn of the way different atoms interact and what new chemicals they may form. The periodic table has proven to be one of science's most powerful predictive tools. Through discovery and seeking patterns, it links together in a single page all material matter. Scientists can read far more information from the table, as an English scholar might when reading Shakespeare. I hope after reading this book you too see more of the awe-inspiring power in the table which is hung simply and modestly in every science classroom around the world.

Some elements have been used by man for centuries, while others are only now finding prominence outside of a laboratory. Modern chemists and material scientists start some of their experimentation out of the lab and inside computers. They use the vast theoretical understanding compiled by previous generations to simulate various chemical reactions. Thousands of experiments can be virtually run in the time it would take a traditional chemist to conduct just a single one. When there is confidence that a reaction might be successful then work moves to the lab. Such research can only accelerate the discovery of new chemicals or materials which find novel fantastic uses.

The acceleration of science seems to continue, pushing the boundaries of our knowledge to ever further horizons. Each element added to the periodic table reveals more about the fabric of nature. With each step we can only guess at what exciting new discoveries might be made, stirring an evolution of science and our society.

Index

Author's Acknowledgements

This project which will, I hope, channel my enthusiasm for a particular strand of science to a broad audience of all ages has only been made possible due to the commitment of the editorial team at SJG Publishing and I am very grateful for their guidance and patience.

Thank you to all my friends and family for understanding why I was a hermit during the writing of the book. I would like to especially thank my wife Emily, who I married on August 5th of this year 2016. Without her patience, encouragement and love it would have been difficult to have worked all those late nights and weekends. I suppose it did give you some time to organise wedding things without me interfering too much!

I would also like to add a dedication to my granddad Bill, William (Bill) James Hollick, who passed away March 18th 2016 aged an amazing 94. One of your many stories made it into the book granddad, along with a photo of you looking rather dapper (page 44).

Picture credits

The photographs on pages 8, 12, 13, 15, 18, 33, 37, 44, 45, 46, 61, 63, 71, 72, 73, 75, 76, 78, 79, 80, 82, 105, 107, 124, 125, 132, 133 (5 images), 138, 144, 148, 152, 158, 159, 162, 165, 166, 167, 169, 170, 172, 173, 175, 177, 178, 179, 180, 181, 183 & 187 are reproduced courtesy of the Science Photo Library.
Page 110 reproduced courtesy of Getty Images.
Pages 34, 51 (2 images) , 52, 54, 59, 60, 62, 68, 69, 70, 83, 84, 89, 93, 96, 97, 98, 103, 104, 109, 116, 119, 120, 127, 128, 131, 135, 145, 146, 147, 149, 150, 154, 157, 163 & 176 are reproduced courtesy of Shutterstock.com.
Pages 43 Zeynel Cebeci at English Language Wikipedia, 55 Rutgers University Libraries at English Language Wikipedia, 130 Fonds Eugene Trutat at English Language Wikipedia, 174 Donald Cooksey at English Language Wikipedia are reproduced courtesy of Wikipedia Commons.
Pages 41 & 49 are reproduced courtesy of the author Ben Still.

Illustration Credits

All diagrams and illustrations kindly provided by Jon Davis

Editorial Director Trevor Davies
Packaged by Susanna Geoghegan
Copy Editor Fiona Thornton
Design Bag of Badgers
Illustrations by Jon Davis
Indexer Jeremy Complin
Production Assistant Manager Lucy Carter